The Channel Tunnel Story

OTHER BOOKS ON TUNNELLING FROM E & FN SPON

Buried Structures: Static and Dynamic Strength
P.S. Bulson

Engineering the Channel Tunnel
Edited by C. Kirkland

Pipejacking and Microtunnelling
J.C. Thomson

Soil Movements Induced by Tunnelling and their Effects on Pipelines and Structures
P.B. Attewell, A.R. Selby and J. Yeates

Soil–Structure Interaction: Numerical Analysis and Modelling
Edited by J.W. Bull

Structural Grouts
Edited by P.L.J. Domone and S.A. Jefferis

The Stability of Slopes
E.N. Bromhead

Tunnelling '94
The Institution of Mining and Metallurgy

Tunnelling Contracts and Site Investigation
P.B. Attewell

Tunnelling: Design, Stability and Construction
B.N. Whittaker and R.C. Firth

Underground Excavations in Rock
E. Hoek and E.T. Brown

The Channel Tunnel Story

Graham Anderson and Ben Roskrow

E & FN SPON
An Imprint of Chapman & Hall
London · Glasgow · Weinheim · New York · Tokyo · Melbourne · Madras

Published by E & F N Spon, an imprint of Chapman & Hall, 2-6 Boundary Row, London SE1 8HN

Chapman & Hall, 2-6 Boundary Row, London SE1 8HN, UK

Blackie Academic & Professional, Wester Cleddens Road, Bishopbriggs, Glasgow G64 2NZ, UK

Chapman & Hall GmbH, Pappelallee 3, 69469 Weinheim, Germany

Chapman & Hall Inc., One Penn Plaza, 41st Floor, New York NY10119, USA

Chapman & Hall Japan, Thomson Publishing Japan, Hirakawacho Nemoto Building, 6F, 1-7-11 Hirakawa-cho, Chiyoda-ku, Tokyo 102, Japan

Chapman & Hall Australia, Thomas Nelson Australia, 102 Dodds Street, South Melbourne, Victoria 3205, Australia

Chapman & Hall India, R. Seshadri, 32 Second Main Road, CIT East, Madras 600 035, India

First edition 1994

Printed in England by Clays Ltd, St Ives plc

ISBN 0 419 19620 X

A catalogue record for this book is available from the British Library.

♾ Printed on permanent acid-free text paper, manufactured in accordance with ANSI/NISO Z39.48–1992 and ANSI/NISO Z39.48–1984 (Permanence of Paper)

To the British and French construction workers
who built the Channel Tunnel

Contents

Preface

We have written *The Channel Tunnel Story* from two distinct perspectives. We make no apology for this, but it does merit a little explanation.

Firstly, we wanted to stress that this was a construction project. This may seem an obvious statement, but with so much of the media coverage focused on the financial and legal aspects of the scheme, we felt there was a danger that the huge achievements of the construction workers – 10 of whom were killed on the project – would be forgotten.

Of course the financial difficulties and contractual rows form a major part of the book. But in a small way we tried to redress the balance back towards the construction process hence why the three interviews contained in the book are all with people from contractor Transmanche Link – former chief executive Jack Lemley, UK tunnelling director John King, and tunnel miner Graham Fagg.

Secondly, the book is written from a British perspective. This is partly because we have been covering the project for the best part of the last decade from Britain. But it also reflects that the Channel Tunnel was a far more controversial project in Britain than in France. The political opposition was stronger, the interest was greater – indeed in some strange way the tunnel seemed to threaten a change in Britain's whole national identity.

In France, things were far less heated. Indeed, there were rumours that an opinion poll in the South of France revealed a significant minority of those questioned who thought there already was a Channel Tunnel!

If there were differences in attitude, in another way the project was a revelation. Despite the stresses and strains, rarely did those working on it divide on nationalistic lines. Contractor and client may have been at each other's throats. British and French rarely were – except when Peter Costain mentioned the rugby.

Perhaps this was a small sign that progress towards a more united Europe is moving faster than any of us realise.

The help we received in writing and researching the book was considerable, and gratefully received. We would like to thank, in particular, *Construction News* for giving us the opportunity to cover the project and allowing us to ransack its back issues and files; *Construction News* editor

John Pullin for his support; Alan Thompson for his assistance with the technology; Ian Crawford for the production; Nick Clarke, Martin Tribe and Alan Peterson at E.&F.N. Spon; but most of all our partners Ros and Tracy, for whom this book created huge amounts of work.

Thanks go to the following who have granted us permission to use their material in this book. The tunnel diagrams are reproduced courtesy of John Hopkins; the maps courtesy of Railway Gazette International*;* Private Eye *cover courtesy of* Private Eye*; Jak cartoons courtesy of Jak and the* Evening Standard*. All photographs are reproduced courtesy of Q.A. Photos of Hythe in Kent except the artist's impression of the Channel Expressway scheme which is reproduced courtesy of Sealink; and the photograph of work on the failed 1970s Channel Tunnel scheme which is reproduced courtesy of Monitor Syndication. We were unable to establish who holds the copyright for the artist's impression of the EuroRoute project.*

We would like to thank Tarmac plc for allowing us to reproduce part of its own lists of key dates and key facts.

The cover photograph is reproduced courtesy of Q.A. Photos.

Introduction

"It is simply a project whose time has come."
A TML manager

On February 17 1990, *The Sun* newspaper carried a photograph of a Channel Tunnel construction worker asleep on one of the tunnel sites. The headline read, "The Chunnel worker-z-z". The caption continued: "Here is more amazing proof that Britain's dozy Channel Tunnel workers spend shifts SNORING instead of BORING." The story – later exposed as a set-up sold to *The Sun* by an office worker on the project – caused a furore on site. The construction workers were risking their lives to make history. Yet most of the time, their efforts were completely ignored. And when the media did sit up and take notice, it was only to run stories such as this.

Whatever the outside world thinks, to almost anyone in the construction industry, the Channel Tunnel is the greatest project since the pyramids. With a final cost of around £10 billion, it is certainly one of the biggest, and the 50,000 or so site workers and office staff who worked on the project at different times – never mind the tens of thousands who helped supply the materials and machinery needed to build it – are unspeakably proud. Their achievement is immense. Yet they feel they have been let down – ridiculed, even – by a British public and media all too ready to whine and complain, even when what they were complaining about was untrue.

Anyone doubting this analysis only has to carry out a "vox pop" among friends and relations or in their local pub. Or alternatively, like us, they could try interesting publishers in a book on the subject.

"You want to write about the opening of the Channel Tunnel. You'll want the fiction department then," joked one helpful editor.

"So this is going to be a sort of Great British disaster story," said another.

The disappointment felt by the site workers and their managers, whose achievement the Channel Tunnel is, is caused not so much by British cynicism: that much was predictable. Some aspects of the project probably merit such treatment. After all, the construction workers know better than most about the actual and near disasters. More, they feel that few people seem to realise the scale of what has been undertaken.

Contrary to popular belief, the Channel Tunnel was not completed with picks and shovels by two teams of cowboy builders with their backsides hanging out of their trousers, with one group smoking Gauloises and the other Old Holborn.

Depending on what you call the start date, it took the best part of a decade to complete. The same amount of money could have built five nuclear power stations or over half a million new homes. Nothing like it has ever been constructed anywhere in the world before. It is also the first land link between Britain and continental Europe since the Ice Age.

And, as if that wasn't enough, it was built with private, not public, money – an experimental arrangement seen by free-market politicians as a crucial acid test of whether private finance could be successfully used for major infrastructure projects.

Cowboy project the Channel Tunnel is not.

But, as anyone who worked on the sites at Folkestone or Sangatte will confirm, that is not to say the project has not faced huge problems, and even tragedies. For a start, it killed ten construction workers, eight on the British side and two on the French, usually in horrific circumstances. One died after being caught between one of the huge segments of concrete tunnel lining and the wall of the tunnel. Another three were crushed by the trains carrying materials into and waste away from the tunnelling faces.

Industrial relations on the project were very good. Despite the safety problems – especially in the early stages of the tunnelling work – there were few disputes. But many times the project teetered on the brink of complete collapse. Without exception, the rows were between the builders and the tunnel's owners Eurotunnel, and were over money and delays – all construction disputes are.

The basis of many of the disputes was simple and was neatly summed up by one of Eurotunnel's top executives. "Just imagine this. You have asked a builder to build you a kitchen. You have agreed a price. But no-one has worked out what exactly is going to be built. So the builder thinks the worktop will be formica. And you think you are getting marble."

The root of the rows may have been simple, but the amount of money at stake was huge – the contractors claimed a total of £1.3 billion extra from Eurotunnel on the mechanical and electrical work alone. And, coupled with delays caused by a patch of unexpectedly wet ground just off the British coast, and what the builders described as late design changes by Eurotunnel, the international consortium of over 200 banks backing the project started getting worried.

Eurotunnel sources muttered about the incompetence of the ten British and French builders and said they were in danger of being sacked. The builders said they were losing money because Eurotunnel was asking them to do things not in the contract and then refusing to pay them for the extra work. So in their turn, they made noises about walking off site and leaving Eurotunnel's top brass to stew in their own juice. There were even rumours that the banks would sack both of them and take over the project directly themselves. None did as they threatened, but when the strong characters of the key players in Eurotunnel and TML are added to the mayhem, it is easy to see why the atmosphere was so electric and so tense. And why the Channel Tunnel kept so many journalists in copy for so long. Indeed, many senior managers – some of whom found themselves working 100 hours plus a week as they tried to keep the project on the rails – today still cannot quite believe that construction is over.

So why is it that the Channel Tunnel is almost complete when everything was stacked against it?

"It is," said one fatalistic TML manager, "simply a project whose time has come."

Another, more pragmatic, reason is that while the directors, politicians and bankers were attacking each other, making headlines and predicting the project's imminent collapse, the construction workers were simply getting on and building it.

Key dates

January 1975
Rail tunnel project abandoned by Labour Government because of cost.

December 1980
Contractor Tarmac and merchant bank Robert Fleming put plans for a new tunnel scheme to the British Government.

Summer 1984
Channel Tunnel Group (CTG) formed by Tarmac, Wimpey, Costain, Balfour Beatty, Taylor Woodrow, Midland Bank and National Westminster Bank.

November 1984
British Prime Minister Margaret Thatcher and French President François Mitterrand back the idea of a fixed cross-Channel link.

April 1985
British and French governments invite bids from private firms to design, build and operate a fixed link

June 1985
CTG teams up with France-Manche, a grouping of five top French contractors and three banks.

October 1985
The deadline for bids to be submitted. Four leading contenders – EuroRoute, Channel Expressway, Eurobridge and the CTG rail tunnel project.

January 1986
Mandate awarded to the Channel Tunnel Group's rail tunnels at a ceremony in Lille, northern France.

February 1986
Treaty of Canterbury is signed allowing the project to go ahead.

April 1987
French Parliament approves the project.

July 1987
Channel Tunnel Bill receives Royal Assent.

November 1987
£5 billion bank loan is agreed and the final tranche of equity raised, worth £750 million. Construction proper starts.

June 1988
First tunnel drive gets under way in France on the land service tunnel.

December 1988
British service tunnel boring machine starts drive towards France.

April 1989
French land service tunnel breaks through – the first tunnel drive to do so.

December 1990
Service tunnel breaks through under the sea linking Britain and France.

November 1992
Undersea border between Britain and France officially marked.

July 1992
First shuttle wagon delivered.

December 1992
First shuttle locomotive delivered.

September 1993
First shuttle locomotive travels through the tunnel.

December 1993
Contractor Transmanche Link hands tunnel over to client Eurotunnel.

May 1994
Channel Tunnel formally opened by the Queen and President Mitterrand.

Mid 1994
Freight and passenger services start operations.

Key facts

■ Before the Channel Tunnel, the construction industry regarded tunnelling rates of 25 metres a day as a great success. But on this project, tunnelling records were broken and re-broken. Rates of 50 metres a day were common. On February 24 1991, the British machine digging the southernmost running tunnel heading towards France completed a huge 75.5 metres.

■ The contractors claim the concrete for the tunnel linings is the strongest ever produced, with a crushing strength of roughly double that used in the pressure vessel of a nuclear power station.

■ The undersea crossover caverns are 156 metres long, 18.1 metres wide and 10.5 metres high.

■ Over 800,000 tunnel linings were installed, weighing from under one tonne to 9 tonnes each.

■ 74 acres of new land was created at the foot of Shakespeare Cliff by using the tunnel spoil.

■ The project used 11 tunnel boring machines. The TBMs that bored the main running tunnels weighed over 2,700 tonnes each.

■ In total, more than 160 kilometres of tunnel have been built.

■ 200,000 trees and shrubs have been planted at the UK terminal.

■ The internal diameter of the running tunnels is 7.6 metres and the service tunnel 4.8 metres. The internal diameter of the Piccadilly Line in London is 3.56 metres.

■ When TML was formed in 1986 it had six staff. At the peak of construction it had 15,000. Daily expenditure averaged over £3 million.

■ The shuttle wagons designed to carry cars and their passengers are the largest railway wagons in the world.

■ The power required for the trains alone is more than 180 megawatts, equivalent to the peak load of a city with a population of 250,000.

■ The French terminal covers a greater area than Heathrow Airport.

■ Orders worth more than £2.2 billion for equipment and services were placed, £1.2 billion on the UK side and £1 billion in France.

■ On the UK side, 92 per cent of orders were placed in the UK. On the French side, 72 per cent of orders were placed in France.

■ The shuttle trains carrying vehicles and passengers take half an hour to cross from Britain to France.

■ The through express "Eurostar" trains to be run by British Rail and SNCF will travel from London to Paris in three hours – less when the UK's high-speed line is built.

■ The tunnel's cooling system is equivalent to 100,000 domestic refridgerators.

■ The contractors installed in the tunnels over 550 kilometres of pipework, 1,300 kilometres of power cables and 20,000 light fittings.

■ The tunnel will use 38 electric locomotives, 270 HGV wagons and 252 tourist wagons.

1 The end of a dream

"The tunnel is not a commercial project. It is a political project . . . to bring votes to the Government of the day while leaving the costs and problems to be faced by others later."

Andrew Alexander, Daily Mail

The story of this project – the latest in a long line of attempts to build a fixed link across the Channel – really begins early in 1975, with the abject failure of another, completely separate, scheme. Two years before, agreement had been reached between the British Prime Minister Edward Heath and the French President Georges Pompidou to build a Channel Tunnel. But within months Pompidou had died, Heath had lost the 1974 general election in the wake of the three-day week and the miners' strike, and new Prime Minister Harold Wilson was desperate to cut out all "unnecessary" spending.

Preparatory construction work had already started and in fact the French had already run into serious technical problems with water pouring into the access tunnel under the town of Sangatte.

Then in January 1975 the Labour government unilaterally brought work to a halt with a brand new £500,000 Nuttall-Priestley tunnelling machine literally hours away from starting to dig the pilot tunnel towards France through the chalk beneath Shakespeare Cliff.

The British politicians were worried about the cost and had other priorities. The Channel Tunnel was not deemed necessary – indeed many in the Labour cabinet were reported to be against it for reasons of traditional British xenophobia – and much to the fury of Paris, that was the end of that.

Well, up to a point. Because although popular opinion was chary of closer ties with continental Europe and the politicians were unenthusiastic, business was acutely aware that trade between Britain and its European neighbours was growing fast – and the time and cost of crossing the Channel put British exporters at a great disadvantage.

PRIVATE EYE

No. 342
Friday
24 Jan. '75

12p

CHUNNEL—
BRITAIN BACKS OUT

Merde alors!
All this way for
nothing

COURTESY OF PRIVATE EYE

In addition, the British construction industry was understandably keen on the idea of several billion pounds' worth of work at a time when public capital spending was on a very tight rein.

The British and French railway networks were also positive, despite a somewhat tense relationship between the two. The French were especially enthusiastic. They were developing their high-speed TGV system and believed they could effectively compete with the airlines if they had a high-speed London to Paris rail route.

By 1979, BR and SNCF had produced plans for a small-bore rail tunnel scheme, nicknamed the mousehole, and that same year the political climate changed fundamentally when Margaret Thatcher decisively won the general election for the Conservatives.

British business still badly wanted a fixed link, as did a work-hungry construction industry. The French were still keen.

So perhaps the new government would look more positively on the idea than its Labour predecessor? It did. But in one respect, Labour and Tories were in complete agreement. Spending any more public money on such a project was out of the question.

The dream of building a Channel Tunnel has nagged away at the consciousness of civil engineers, entrepreneurs and military experts since the early nineteenth century. In 1802, French engineer Albert Mathieu Favier gave Napoleon details of a scheme involving two tunnels, one which would be candle-lit for horse-drawn carriages and the second acting as a sort of drain to take away water seeping into it.

In the 1870s construction actually started on the British and French coasts with a view to building a railway tunnel between Folkestone and Cap Gris-Nez. But the British, as they did 100 years later, called a halt, worried about the security implications.

Despite the Labour government's decision to pull out in 1975, supporters of the project kept plugging away and it remained in the headlines.

In December 1980, Tarmac and merchant bank Robert Fleming sent the Government detailed plans for dual rail tunnels plus a smaller service tunnel similar to that now being built. The British government had agreed to release to Tarmac the drawings and designs of the 1975 scheme. Having studied them, the contractor argued that with the help of private finance, construction of a similar project could start "in the shortest time".

The funding problems meant that Tarmac opted for building it in three phases. The first phase, costing £830 million, would consist of a single 7 metre diameter running tunnel and a 4.5 metre diameter service tunnel. There would be no terminals and the tunnel at this stage would cater for through rail traffic only.

Phase two would involve building terminals at Cheriton, near Folkestone, and Frethun, and providing shuttles to take vehicles through the tunnel. This would cost £420 million.

The final phase, costing a further £480 million, would involve boring an extra running tunnel and expanding the terminal facilities.

The rail-link option received a further boost in February 1981, when the House of Commons Transport Committee called for legislation to be put before Parliament to allow a Channel Tunnel project to go ahead.

The following year, a joint study group set up by the British and French governments concluded that "the balance of advantage lies with bored twin tunnels with a vehicle shuttle constructed if necessary in phases".

Then, in 1982, two British and three French banks – Midland, National Westminster, Banque Indosuez, Banque Nationale de Paris and Crédit Lyonnais – formed the Franco British Channel Link Financing Group. They were responding to the French and British governments' worries about the likely cost of the project, should it be restarted, and in particular the British government's desire for it to be privately financed.

Their work took them over one and a half years to complete and, with hindsight, gave the idea of a fixed link a crucial boost – although when first published early in 1984 many commentators felt it meant the end of the project.

The report spelt out the action that the British and French governments would have to take if the financial markets were to give a Channel link sufficient backing – advice which in many cases ministers later followed. For example, they would have to set out in as much detail as possible the type of fixed link required, the nature of the owning entity and the terms of the operating licence on offer.

In addition, the report gave supporters of the twin rail tunnel project – rather than a link involving road tunnels or bridges, or a combination of both – a huge boost by declaring that "this option is most likely to be financed by the capital markets".

It added: "One factor of critical importance in determining banking acceptability was the level of technical and construction risk, with its consequent impact on total construction cost."

"Having consulted various experts, it was concluded that the technology associated with bored tunnels was well proven, even though there would still be the risk that geological problems might exist beyond those initially identified, leading to possible delays and additional costs.

"In contrast, the bridge and composite schemes rely on relatively untried technology and raise unresolved questions concerning navigational safety."

But, having supported the idea of a rail link, the banks then threw a spanner in the works. The report confirmed that the project could be funded by private money and the banks proposed two financing methods, but both involved the two governments taking at least part of the risk and offering various forms of support and financial guarantees, especially in the event of substantial cost overruns. Such proposals might have been acceptable to the socialist French president, François Mitterrand, but they were certainly was not to the liking of Margaret Thatcher or her Transport Secretary, the right-wing free-marketeer Nicholas Ridley.

The banks recognised that this conclusion was likely to be unpopular in Westminster and tried to soften the blow. They said at the time: "It is accepted that neither financing structure completely meets the original requirements of the governments.

"However, within both structures, the objective remains to contain the level of government support for the project and to demonstrate that genuine risk-sharing arrangements are possible."

It was an approach that was way ahead of its time – at least as far as the British government was concerned. After all, such risk-sharing arrangements have still not been developed to allow construction work to start on the high-speed rail link between London and the Channel Tunnel, almost a decade later.

Certainly, Ridley was not about to alter his approach to the project on the basis of one report, however weighty. He told the House of Commons on the day it was published: "It has been and remains the government's firm position that any project would have to be financed entirely without the assistance of public funds and without commercial guarantees by the government.

"So far, we have seen no proposal which demonstrates that it can meet this condition."

For the media, this represented a major setback. "Channel link hopes dashed" read the front-page headline in the industry newspaper *Construction News* that week. The *Financial Times* was more cautious, but not by much. "Banks' chunnel scheme rebuffed", it said.

Long-time construction journalist John Allen, the editor of *Construction News*, wrote: "This is patently a work for which governments ought to take funding responsibility. The fact that they refuse on the grounds of misguided economic policies only puts its realisation further into the future."

But Nicholas Ridley was careful not to close the door on the project completely. He also told MPs: "The government remains willing to consider facilitating a fixed link, in collaboration with the French government, provided that the necessary financial, technical and other aspects are satisfactorily dealt with." He and the Prime Minister clearly still believed that the private sector would be prepared to take on the project and all associated risks. Within months they were to be proved right.

The bank's report was published in May 1984. By October, the Channel Tunnel Group – a loose grouping of British contractors proposing the increasingly popular rail link – announced that they could build a Channel Tunnel with no government financial help at all. CTG said they had carried out a new study and, with the backing of independent consultants and interested banks, had cut their construction programme from six years to four and a half. "This will significantly reduce the total amount of funding required and make the project even more attractive to private investment," said CTG stalwart and Wimpey director Tony Gueterbock. "CTG now claim that they can fully comply with the British government's criteria for the construction of a fixed cross-Channel link," he added. "We have submitted our new proposals to the Department of Transport and I hope that this will release the log-jam of indecision which has inhibited progress on the project since the publication of the financing group's report last May."

In fact, talk of a "log-jam of indecision" was somewhat unfair. Certainly the banks' report and Nicholas Ridley's response had caused all supporters of the project to pause for thought, but by now, problems or not, the idea of building a Channel link had huge momentum behind it, enough to keep it going for the moment at least. Public interest was growing fast and ministers in London and Paris had kept in close touch. But the time had now come for the two governments to agree a common approach. If that could not be achieved, then all the efforts to date would have been a waste of time.

The key political breakthrough came at a regular meeting between Thatcher and Mitterrand in Paris in November 1984. Following their talks they issued a joint communiqué saying that both governments regarded the building of a fixed link across the Channel as being in their mutual interests. The only difference between them was over how it should be financed. The French feared that the project might never be completed without at least some financial support from the public purse. But, as she had made repeatedly clear, Margaret Thatcher wanted the whole thing to be a symbol of what the private sector could achieve on its own. She won.

Such positive political noises were more than supporters of the project had dared to hope for. At long last, the race was under starter's orders, and if it could be privately financed it should escape the uncertainties of public spending and Treasury control.

But why was there political support for the idea of a cross-Channel link at all?

Writing in the *Daily Mail* years later, the paper's City Editor Andrew Alexander summed up the feelings of many sceptics: "The tunnel is not a commercial project. It is a political project – just like the Humber Bridge – to bring votes to the government of the day while leaving the costs and problems to be faced by others later," he wrote. "For those who regard all this as pessimistic and even cynical, let me point out that the benefits to the public of the tunnel...are very limited. Ferries are labour-intensive, while a fixed link is capital-intensive. This means that the tunnel needs above all to boost its traffic. Thus it would pay those operating the tunnel to put the ferries out of business in a price war. Then, when the ferries are almost extinct (and many more jobs have been lost than the tunnel can ever provide) prices would be allowed to drift up again and the tunnel operators would be rolling in money – and you would be paying more to cross than if the ferries were still operating."

He concluded: "And all this would be just to make a marginal cut in the time it takes you from leaving home to reach a European destination. The fact that the hovercraft service, which provides the swiftest journey across the Channel, is unable to operate profitably is a warning to those who think that the travelling public is frantically looking for faster crossings."

But despite the cynics, the fixed link had political support in abundance. It is clear why President Mitterrand should have taken to the idea. Big construction projects – his "grands projets" – have littered his presidential terms of office. Perhaps he sees them as his bequest to the nation, marking his time in power. But there were pragmatic reasons as well. The north of France was like the north of Britain – reliant on heavy industries that were suffering badly and with high unemployment. With luck, not only would the Channel Tunnel create construction jobs, it would give the whole Nord-Pas de Calais region a much-needed boost. Indeed, several UK property experts believed that with room for development so tight in Kent and across the south-east of England, many British firms would relocate to northern France, where land was readily available and rents and labour were relatively cheap.

French businesses, like their British counterparts, wanted the link, and in France the policies of politicians and the needs of industry have always been more closely aligned than on the northern side of the water.

Once, several years later, when the project had hit problems, one of the leading French contractors apparently suggested they should tell their political masters of their difficulties, both to ask advice and because of the possibility of political embarrassment should the scheme fail. He was given a reception with President Mitterrand within days. His British counterparts complained they had to wait several months, and then only got to see Transport Secretary Cecil Parkinson, rather than Margaret Thatcher.

If Mitterrand's motives for supporting the Channel Tunnel seemed straightforward, Thatcher's were less clear. She would undoubtedly have been lobbied hard by business and construction interests supporting the idea of a fixed link, but, as we now know, she was not one to be swayed easily by the voices of vested interest. In addition, her lukewarm attitude even then to much of what was emanating from Brussels in the name of European unity, plus the "little Englander" elements on the right of her own party – where she had her political power base – made her enthusiasm even more surprising, at least to outside observers. Some suspected that her motives were in fact very similar to those of President Mitterrand – a desire to leave a monument to her time in power. But, whatever the reason for it, Thatcher's backing meant that many of her more sceptical ministers – Nicholas Ridley was one – remained neutral rather than opposing the idea outright.

Journalists who covered the 1984 meeting with President Mitterrand said they felt that Thatcher's support was a gift to the French leader, who was strongly in favour of the project.

The Guardian's Paris correspondent Paul Webster wrote in the edition of December 1 1984, that the agreement "seemed a personal favour to President Mitterrand whom Mrs Thatcher repeatedly praised for solving Britain's budget demands earlier this year."

He added: "Pressure for a rapid agreement on the tunnel came mainly from France. Only the night before the joint press conference at the Elysée, Mrs Thatcher was sceptical about a Channel link. The French, however, believe it can be under way before the end of the decade."

Sir Nicholas Henderson, the former British Ambassador to Paris, a keen supporter of the project and later to become chairman of the Channel Tunnel Group, was someone very close to the centre of the action. He later wrote: "The Prime Minister's decision had two main impulses: the desire to encourage some major industrial enterprise such as Britain had not carried out since the Second World War and one which would provide a good deal of employment; and the wish to make some positive move towards our European Community partners with whom we had been wrangling for so long over the Community budget."[1] But he added: "Months later those close to the Prime Minister continued to be surprised by her decision to commit herself so wholeheartedly to the project."

Whatever the reason for it, the Channel Tunnel would not have been built without the political impetus Margaret Thatcher's support gave it.

By now, firms interested in building or financing the Channel Tunnel had started to organise themselves around specific proposals.

The EuroRoute, Margaret Thatcher's personal favourite, was a drive-through scheme costing up to £5 billion. Its backers included several of the big names of British industry – British Steel, Trafalgar House, British Shipbuilders and Fairclough.

It involved building two artificial islands in the Channel. The islands would be linked to the shore by bridges and to each other by a tunnel under the central shipping lanes.

A second group was investigating a huge suspension bridge (later described by a disbelieving Department of Transport official as a project "for the next century rather than this one").

There was also the Channel Tunnel Group and its plan for a dual rail tunnel separated by a smaller service and emergency exit tunnel. Its founder members were five of the biggest construction firms in the UK – two of the contractors from the defunct 1975 scheme, Taylor Woodrow and Balfour Beatty, plus Costain, Wimpey and Tarmac. Within months, they were to be joined by all five of the banks who had written the financial report on the project which had come down in favour of the rail link.

This was already the experts' clear favourite. It had the backing of the banks, government officials and most of the British and French construction industries. Millions of pounds had already been spent by the interested firms on assessments, lobbying and basic design work. But no-one was anywhere near building anything yet, and in late 1984, the prospect of the contractors turning the first sod still seemed an awful long way off, political backing or no. For a start, the British government had made it repeatedly clear that it was not going to put a penny directly into the Channel Tunnel. And, whatever the potential promoters said about being able to finance it entirely within the private sector, they would still have to find someone to put up the money.

By Christmas 1984, the Channel Tunnel had almost got to first base. But that was all. Only the fanatics, and possibly President Mitterrand who was reported to be prepared to back it with public money if necessary, as yet really believed it would be built. As one cynical civil servant said: "You have to remember that a lot of us had been here ten years before."

[1] From *Channels and Tunnels*, by Sir Nicholas Henderson, published by Weidenfeld and Nicolson, 1987

2 A *fait accompli*?

> *"I cannot yet tell whether a fixed link will be built across the Channel or not. What I can say is that the private sector now has a unique opportunity."*
>
> Nicholas Ridley MP, Secretary of State for Transport

As soon as Christmas 1984 was over, it became clear that both the British and French governments were committed to pushing the project forward – so much so that in Westminster a growing number of backbench Tories were becoming restless at what they saw as ministers' attempts to railroad approval through Parliament and the planning process with minimal consultation.

The ferry companies and the inhabitants of Kent, especially Dover and Folkestone, began to realise that the idea of a Channel crossing was again being taken very seriously and began lobbying to protect their interests.

If the link was to progress as a privately financed project, the first step was for both governments to define what they wanted from the private construction firms and financiers interested in building and running it. To that end an official Anglo-French working party of civil servants was created. Their role was to set out the technical, safety and financial requirements that any cross-Channel link would have to satisfy. They would consider how long the private sector should be allowed to operate the link for, and whether it should eventually be transferred to public control; how safe it would need to be to provide the public with reasonable protection in case of accidents or fires; the implications for national security; and, in the case of a bridge, navigation issues in the English Channel: what would happen if the bridge was hit by a supertanker, for example?

They were also to look at the political guarantees that the two governments could offer the scheme's promoters to stop politicians in London or Paris trying to interfere with fare levels or, even worse, trying to halt the project after construction work had started, as had happened in 1975.

The work of the civil servants was predictable. Such issues had to be addressed, and ministers had to decide exactly what it was they wanted the private sector to do. What was less predictable was the timetable. The working party held its first meeting in the second week of January 1985 and was told by ministers to complete its deliberations by the end of February. If there were no unexpected complications, and if Anglo-French relations remained good, the two governments hoped to invite private sector promoters to bid to design, finance, construct and operate a fixed link just a month later.

This sudden sense of urgency again underlined the British and French governments' fast-growing commitment – a commitment increasingly underpinned by the realisation on both sides of the Channel that the project could be exploited for good old-fashioned party political gain. Elections were on the horizon in both countries. The French parliamentary elections were due to take place in the spring of 1986. And in 1988 there was due to be a French presidential election plus a British general election (although in the event Mrs Thatcher went to the country early, in June 1987).

The two leaders might have been poles apart politically, but both recognised the great potential electoral benefits of the Channel link because of its high profile, futuristic image and its job-creating potential.

It was rumoured that Margaret Thatcher had promised President Mitterrand that the winning scheme would be announced in time to benefit the socialists in their French parliamentary elections.

If that deadline was met, both the British Conservatives and the French President should then benefit from the sight of construction work starting before voters went to the polls again in 1988. All of a sudden, the atmosphere had changed. Political realities had begun to intrude on what until now had been a rather hypothetical debate.

According to Sir Nicholas Henderson, who was to become chairman of the Channel Tunnel group in February 1985, the tight deadline did have some beneficial effects, however: "This time pressure led to the establishment in Whitehall of a remarkably effective organisation for tackling the wide-ranging problems, a system which forestalled interdepartmental wrangles and continued as the main motor for moving the British side along until the mandate was awarded." [1] Nonetheless, the timetable was beginning to look very tight indeed.

By April 2 1985 that timetable was looking even tighter.

Transport Secretary Nicholas Ridley made a statement in the House of Commons announcing publication of the "Invitation to Promoters" as planned. This consisted of the guidelines future operators of the Channel link would have to comply with if their schemes were to be considered seriously by ministers.

The document was divided into four main parts, covering the role of the governments; the legal regime under which the link would operate;

finance, and in particular the need for the schemes to demonstrate their financial robustness; and the physical details of the project. This last point involved the promoters producing designs, technical evaluations, reports on the effects of their schemes on jobs and the environment, and how the safety of the public would be ensured. The governments also demanded to know how the schemes would be protected against terrorist attacks. And they said they would want to be sure that whatever scheme was chosen would have sufficient financial backing to be able to deal with cost overruns and construction delays. On top of this they said that promoters would be required to pay deposits worth 150,000 Ecus in both Britain and France when they submitted their plans to the governments, although unsuccessful bidders would have this money refunded. Finally, all this had to be completed by October 31, less than seven months away and a deadline two months earlier than most had anticipated, so that a winner could be announced the following January.

All this was a far cry from the vagaries and uncertainties of the first half of 1984. There was now no doubt that the political will existed in government to see a link built. But could the money be found to fund it?

The sense of urgency was underlined by Transport Secretary Nicholas Ridley when he announced publication of the Invitation to Promoters in the House of Commons. He confirmed that the civil servants' report had been delivered, bang on time, on February 28, and that on March 20 1985 British and French ministers had met to agree the next steps with the intention of reaching a decision on which proposal, if any, should be given the go-ahead.

He added: "We also agreed that we should begin contingency work now on those elements of the Treaty which would be common to any form of link chosen." But Ridley's political antennae were also telling him to beware. In theory no decision had yet been taken on whether there should be a cross-Channel link at all, but already the debate seemed to be about what *sort* of link should be chosen. It was beginning to look like a *fait accompli*, and a number of MPs of all parties did not like it.

Ridley tried to take some of the heat out of the situation: "I do not want honourable members to think that we have prejudged the issues. When they have had time to study the guidelines they will recognise my concern to ensure that there is adequate public consultation, that environmental, social and employment impacts are fully appreciated and that the financial conditions are fully met." He concluded: "I cannot yet tell whether a fixed link will be built across the Channel or not. What I can say is that the private sector now has a unique opportunity. We have reached full agreement with the French on the conditions which the promoters must meet."

If the intense political activity of the first three months of 1985 galvanised the supporters of a Channel link into frenetic action, it did the same for their opponents. James Sherwood, chairman of Sea Containers, which had only bought the ferry company Sealink the previous July for £66 million, set the tone. He was furious. He announced that he would actively oppose the construction of any kind of fixed link and warned that if it went ahead it would hit the existing ferry companies hard, would lead to thousands of job losses and would turn Dover into a ghost town.

The speed with which the project had accelerated early in 1985 had caught many people off guard, but by Easter that year many others were beginning to voice their misgivings, both about the wisdom of building a link at all and about what they saw as the lack of any proper consultation with those affected or consideration of the issues raised.

At the end of April 1985, an anti-link lobby group had been formed to provide a focal point for these concerns and to increase pressure on ministers. Called Flexilink, it was created by the Dover Harbour Board and its members included ports, shipping agents and the Calais and Boulogne Chambers of Commerce as well as the ferry, hovercraft and jetfoil operators. The organisation described itself as "a group of people who normally compete fiercely with each other but who have joined together to raise vital questions about the future of cross-Channel services." It added that its members were "deeply concerned that no real debate about the development of cross-Channel links is taking place and that key issues affecting public welfare and safety are being ignored."

In the debate about a single fixed link, the advantages of the existing services and their capacity to expand and develop was not being considered, argued Flexilink chairman Jonathan Sloggett of the Dover Harbour Board. "The problem with crossing the Channel is not technological but procedural. Formalities, especially for freight, such as quotas, licences and levies, collection of statistics, payment of value-added tax on goods, all of these must be carried out," he said. "If faster cross-Channel traffic is the aim, it would be much more effective, as well as less costly, to streamline these formalities...the fixed link is a solution looking for a problem."

Flexilink's arrival on the scene was greeted with a predictable storm of derision from those competing for the job of building the link, in particular the Channel Tunnel Group. "The announcement by the Dover Harbour Board and their new Flexilink consortium that they will oppose either a Channel Tunnel or a bridge comes as no surprise," said CTG's Tony Gueterbock. "'Anything you can do, we can do better' is their favourite theme song whenever their monopoly hold over the most expensive sea ferry route in the world is threatened," he added.

"It was heard loud and strong during the 1970s when the Channel Tunnel appeared to be a competitive challenge to their future, but when the threat of a Channel Tunnel scheme faded, so did the attractive inducement of reduced tariffs to users."

But concern, especially at the lack of adequate consultation, was growing, and not just in Flexilink's office and in Dover – where the vast majority of the town's population depended directly or indirectly on the ferry trade for their livelihoods.

Across the Channel in Calais local people claimed that everyone was against the project except the socialists, and they only backed it because they were under orders. According to the director of the town's Chamber of Commerce, Guy Flamengt, one-third of Calais' wage earners could lose their jobs if the link was built, in what would be an "economic catastrophe".

Opposition was also growing throughout Kent, with local people worried about how the huge increase in cross-Channel traffic that EuroRoute and the Channel Tunnel Group were predicting (and which they were using to justify building the link at all) would manage to pass through the already congested towns and villages en route to the coast. Strangely enough, they were also worried about planning blight resulting from yet another protracted debate about a high-speed rail link between London and the Channel coast (the cost of such a link was one of the reasons for the abandonment of the 1975 tunnel). As it became clear that the link would sidestep the normal planning regulations, so the rows grew.

In theory, the governments could still decide not to pick any of the schemes presented to them by October 31 1985, but by this stage few people regarded that as a serious option. (British Transport Minister David Mitchell was given the job of holding consultation meetings in Kent. He insisted the government was not presenting the public with a *fait accompli.* It was an uphill struggle – few believed him.)

The most likely course of events was that Mrs Thatcher and President Mitterrand would announce the winning project in January 1986. In Britain, the announcement would be followed by publication of a White Paper and the signing of a Treaty between Britain and France to allow for the construction of the link. After that a Bill giving approval to the project and giving the winning promoters the necessary powers to proceed would be presented to Parliament. All being well, construction would start in 1987.

This process ruled out the usual public enquiry method of considering controversial or significant projects, and meant that comments from outside Parliament would have to be made through representations to the committees that would be studying the Bill in detail. This made even some supporters of the link uneasy.

Groups such as the Royal Town Planning Institute, the Campaign for the Protection of Rural England and lobbyists Transport 2000 all called on the British government to think again, and to hold an enquiry to allow public concerns to be fully aired.

Transport 2000 were not alone in describing the arrangements for pub-

(Top) A tunnelling machine built by Robert L. Priestley arrives on site in Dover to start the unsuccessful 1970s tunnel attempt.
(Right) Not everyone likes the idea.
(Bottom) Margaret Thatcher and French president Francois Mitterrand ratify the treaty in the Elysée Palace in July 1987.

(Top) Sir Nigel Broackes's EuroRoute was hugely ambitious and very expensive, but you could drive across it.
(Bottom) The main backer of the Channel Expressway tunnel project was James Sherwood's Sealink ferry company.

(Top) The bottom of the access shaft at Sangatte in France showing the portals from where the tunnelling machines started their journeys.
(Bottom) The second access tunnel at the foot of Shakespeare Cliff under construction at the end of 1987.

(Top) Building the lagoon at the foot of Shakespeare Cliff for the tunnel spoil.
(Bottom) Engineers position the cutting face of the massive boring machine that drove the UK north running tunnel.

lic consultation as grossly inadequate and without precedent in modern times. "We are at a loss to understand how Mr Ridley imagines that the public can come up with informed views on the relative merits of vast schemes when the only information it has on the schemes comes from the promoters themselves, and when there is no information at all on the effect on the country itself .... These issues are too important to be left to the promoters and the government." The lobbyists added: "The speed with which this vast issue is being dealt with is astonishing. It is not cynical to link the haste with the election programmes of the two countries involved, and we protest against the surrender of proper democratic process to electoral advantage.

"The government must hold a public enquiry into the schemes. The hybrid bill procedure is most unlikely to satisfy the public need for information, neither will it assure people of the reasonableness of the decision."

But Transport Secretary Nicholas Ridley stuck to his guns. A public enquiry would cause long delays and would mean the project becoming a political ping-pong ball in the run-up to elections in Britain and France, and the delays could frighten off the private sector financiers. And as the government had said on many occasions, without them there would be no link.

But opposition was not just focused on the lack of consultation and the public's democratic right to know what was going on.

Flexilink and others were beginning successfully to undermine the arguments that the link would create jobs. The ports and the ferry firms said that, yes, short-term construction jobs would of course be created, but in the medium and long term the project would have quite the reverse effect. And dramatically so. Their concern was that coexistence between existing ferry services and a fixed link would be impossible. They detailed their arguments in a paper presented by Flexilink to the House of Commons Transport Committee in the autumn of 1985. "The consequences of the ferries having to compete with a fixed link are far-reaching and significant for the community as a whole because the regional and longer short-sea routes would be forced out of business," the report said. "The principal reason for this is that following the destabilising of the market resulting from the significant increase in capacity when a fixed link opened, tariffs on the Dover to Calais route would fall progressively, as would tariffs on other routes. Thus all ferry tariffs would quickly reach a point at which it was no longer possible for the industry to give financial support derived from the financial security of the Dover to Calais services to regional routes and longer short sea routes."

The report concluded: "As a result the ferries operating on these other routes would be forced out of business and the services would cease." And it added: "Thus the consequence of the construction of a fixed link would be to limit choice rather than increase it for the community benefit."

Flexilink calculated that opening a fixed link would cost anything between 20,000 and 40,000 jobs on the ferries and in the ports, with 60 per cent of those losses occurring in the United Kingdom. It hammered home these points in an aggressive advertising campaign, with slogans such as "The Channel Tunnel – the black hole that will put Britain in the red" and "Hired by the Channel Tunnel: 3,500 employees. Fired by the Channel Tunnel: 40,000 employees".

Opposition voices were becoming louder in Westminster, too. Among the loudest was a group of backbenchers who included Jonathan Aitken and Teddy Taylor, and called themselves "Conservatives against the Tunnel". At local level, Gwladys Payne, the then Labour leader of Dover district council, gave a glimpse of the bitterness felt by many when she criticised the "climate of euphoria" surrounding the possibility of building a fixed link, which meant that "scant attention has been paid to the economic pitfalls which will be generated in its wake, and more particularly to the consequences for the ferry services". She also feared huge job losses and remarked with irony that the British and French governments had now started to look at ways of streamlining customs and immigration procedures and how to improve the infrastructure in Kent to meet the needs of the fixed link, when the council had been pressing for such measures for years to improve the efficiency of crossing the Channel by ferry.

Why could such measures not have been discussed before?

Speaking just after the publication of Flexilink's report for MPs, she said: "Dover is still waiting for the A2 to be upgraded to dual carriageway throughout its length, and is still waiting to hear that the M20 extension to Dover will go ahead as well as urgently required faster and improved rail services to what is Britain's most important passenger and freight port."

Implicit in Gwladys Payne's remarks is the question "why don't we just expand our port facilities rather than spend years and billions of pounds on such a risky venture?".

Others asked the same question, including the economist Sir Alec Cairncross and Sir Alfred Sherman, a strong supporter of Margaret Thatcher and co-founder of the right-wing think-tank, the Centre for Policy Studies. In a pamphlet for the Selsdon Group, a group of free-market Conservatives, Sir Alfred wrote: "The corridor from the Channel ports to the midlands and the north is the most congested in the country. It does not make good sense in terms of economic geography for an island whose main areas of population and economic activity lie close to the coast and estuaries to concentrate traffic from various parts of Europe through the narrow and congested sea crossing and an equally narrow and congested land corridor."

He added: "Would it not make more sense to tackle the inefficiencies of our ports, and in some cases the inadequate access, than to invest billions in an exercise which can only generate further congestion?"

One of the country's top planners, Sir Colin Buchanan, had other concerns. He felt the link could hinder rather than help Britain's trade. With land so expensive in Kent compared with northern France and with development controls across all of southern England so tight, Sir Colin argued that all of the Channel-link-related growth could well take place on the French side of the water, a trend accentuated by UK firms choosing to move there and commute over from Kent. "That could lead to this island becoming the offshore backyard of Europe...Britain the bed and breakfast – is that the way we should be heading?"

Throughout 1985, opponents of a Channel link became increasingly successful at making headlines and putting their concerns across through the media. But although they could make waves, the political realities were such that there was little they could do to force either government to change direction. For a start, the project had the unequivocal support of two strong leaders, Mitterrand and Thatcher. They would not easily change their minds. In addition, the British government had a large majority in the House of Commons and in any case a link of some sort had the support of an overwhelming majority of MPs of all parties – at least if the opinion polls were to be believed.

Support amongst the French, who have a greater liking for "grands projets" than their British neighbours, was if anything even more solid, despite the opposition in Calais and Boulogne. There was also, surprisingly, little pressure on British MPs from the public to oppose a link, except on those representing ferry port constituencies. For public opinion country-wide was not in general anti-link, despite some reservations (although an opinion poll in the *Daily Telegraph* in August 1985 suggested that backing for the idea had waned since 1963, when a similar exercise had been carried out, with support falling from 69 per cent in favour to 50 per cent, and opposition rising to 37 per cent from 17 per cent).

There was little for the project's opponents to do other than to hope for some major stroke of luck that would change everything. But it was unclear how that might come about – even if they won the arguments. Indeed, despite the media stories at the time, one has the feeling today that the various consortia competing for the mandate to build the link took little notice of those who were opposing the whole idea. In Eurotunnel's book telling the story of the project up to breakthrough – when the British and French tunnellers met for the first time under the Channel – the opposition group Flexilink did not merit a single mention. And in former Channel Tunnel Group chairman Sir Nicholas Henderson's account of events leading up to the award of the mandate, he merely mentions their efforts in passing: "Flexilink, representing the ferry industry, the Dover Harbour Board and all those who were against any fixed link, launched a bluff advertising campaign...I am not sure that the slogans did us much harm, though they were grist to the mill of those who accused us of being laggards in public relations. Fortunately we came to an understanding with Sir Nigel Broackes [of EuroRoute] to avoid engaging in an advertising war: This saved both our organisations a lot of money." Hardly the words of a man concerned that the fixed link's opponents might succeed in scuppering the project entirely.

CTG did, however, take care to respond to the claims that a Channel Tunnel would costs tens of thousands of jobs. For a start, they knew that job creation was one of the key reasons the project was receiving such strong political support in London and Paris. Secondly, they believed that, alone amongst the contenders, the rail link proposal would compete with the ferries but not put them out of business. During construction, CTG said it would create 4,500 jobs in East Kent, 70 per cent of which would be filled by local people. When the tunnel services were operating, it calculated it would create between 3,500 and 5,500 jobs. "These figures take account of some possible adverse effects on port-related employment, although the extent of this will, of course, depend on how the ports themselves respond to the challenge of competition."

The demands of the October 31 deadline set in April 1985 by the two governments for plans to be submitted now meant that the time for vague big ideas was over. Detail was called for and those contractors and financiers with serious hopes of building and running a link stepped up their efforts. With the political signals apparently stuck on green, the potential promoters were not unduly worried about the anti-link lobby. They were far more worried about each other.

[1] From *Channels and Tunnels*, by Sir Nicholas Henderson, published by Weidenfeld and Nicolson, 1987

3 The race for the mandate

"It is really a very simple job. It is just very, very big."
A tunneller talking about the CTG project

When Nicholas Ridley announced details of the competition in April 1985, there were two clear front runners, the rail tunnels proposed by the Channel Tunnel Group and the drive-through bridge and tunnel scheme proposed by EuroRoute, a consortium led by Trafalgar House and British Steel.

Despite the disdain poured on its plans by CTG supporters – and despite the opposition of the 1984 banking report – EuroRoute remained a serious contender, partly at least because of widespread misunderstandings about the CTG plans. But EuroRoute also had top-level political support, particularly in the UK. Mrs Thatcher's preference for a drive-though scheme meant it was her choice. Her Trade and Industry Secretary Lord Young also thought it the best option, because it would create the most jobs and because he was worried that a rail-only scheme would present the rail unions with too much power.

At this stage, the French authorities seemed to be favouring the CTG rail link, but as the months passed it became clear that President Mitterrand, like Mrs Thatcher, was studying the EuroRoute option with increasing interest.

The EuroRoute scheme was the most expensive and ambitious of the serious contenders. The driving force behind it was Ian MacGregor, who started working on it when he was chairman of the British Steel Corporation. It was popular with the politicians and the general public and was the media's favourite. But many financial and construction experts had different views. They feared that the high cost gave little room for cost overruns, and they regarded it as a huge technological challenge – perhaps too much of one.

The scheme was certainly awe-inspiring. Driving from the British coast, vehicles would cross a four-lane (two in each direction) suspension bridge to an artificial island 8.5 km out to sea.

At the island, the road descended in a spiral into a 21 km long immersed tube tunnel under the main shipping lanes before re-emerging via another artificial island 7.5 km from the French coast onto another suspension bridge to the shore.

Alongside the road link was a coast-to-coast rail tunnel allowing through passenger trains and freight to travel directly from the United Kingdom to the rest of Europe – giving the scheme an additional income stream from the railways and undermining one of the pluses of the rival Channel Tunnel Group's rail-only link, namely the opportunity to jump on a train in London and get off in Paris three hours later without having left your seat.

The construction cost of the EuroRoute scheme was huge – around £5,000 million at 1985 prices. Interest charges, inflation and provisions would push that up towards £9,000 million. But the advantages were huge too: customers could simply drive across to France. And EuroRoute would create a lot of work.

Ambitious it certainly was, but it had the active backing of some of Europe's biggest industrial and financial firms and it was a truly Anglo-French project (this last point was to be seen as increasingly important as the year passed and the French government made it clear that it would find it very difficult politically to award the mandate to a group with little or no French membership).

British consortium members included Trafalgar House, British Steel, British Shipbuilders, Barclays Bank, Kleinwort Benson, GEC, British Telecom, John Howard, Amec and Associated British Ports. On the French side there was GTM Entrepose, Compagnie Générale d'Electricité, Société Générale, Usinor, Banque Paribas and Alsthom.

By the time the competition for the mandate was officially launched in April 1985, the chairman of the British arm of EuroRoute was Sir Nigel Broackes, the chairman of Trafalgar House, who had succeeded Ian MacGregor in December the previous year. As always, Sir Nigel oozed self-assurance. "EuroRoute will become the symbol that really got industrial Britain moving again, just as the completion of the Queen Mary symbolised the end of the depression of the thirties," he told journalists. He added: "It is a venture to stir the imagination and pride of every one of us."

That EuroRoute was still a serious contender for the mandate at this stage was no small triumph itself. The banks' report, *Finance for a fixed Channel Link*, published in 1984, came down heavily in favour of the Channel Tunnel Group's bored rail tunnels as the only option to be both technically and financially viable. The report said that by the date of opening the basic construction cost, inclusive of inflation and capitalised interest, of a bridge and tunnel composite link could be as much as £24.8 billion, over three times the comparable figure for a twin-bore rail tunnel. The banks also said that the project relied on "relatively untried technology", and as a result it concluded: "Although the drive-through schemes produce the highest revenue...the technical risk and magnitude of financial commitment were beyond acceptability to the financial markets."

The report drew a stinging response from Ian MacGregor, at the time still EuroRoute chairman, who accused the banks of a "serious misjudgement" by ignoring the views of some of the UK's most experienced engineers, and EuroRoute's chief executive Ken Groves added angrily that the banks had relied too heavily on a civil servants' report dated June 1982, "prepared before the full design advantages of EuroRoute had been understood". He said: "This led the banks to ignore the results of an extensive costing exercise, carried out in 1982-3 by consulting engineers Mott, Hay and Anderson. These findings were lodged with the banks in March 1983 and showed that EuroRoute's capital cost would lie between £4.06 and £4.40 billion. The banks had used a sum 50 per cent higher at £6.1 billion without discussing this change or the reasons for it with the promoters of the EuroRoute project."

But EuroRoute could do more than pour cold water on the calculations of its detractors. At a time when unemployment was a sensitive political issue on both sides of the Channel, it offered jobs. Up to 100,000 of them. And those jobs could be spread widely around the two countries. This was a particular advantage in the UK. The idea of creating as many jobs as possible in the depressed areas of the north – rather than in the relatively affluent south-east – was quickly seized upon by both civil servants and ministers. The jobs could be spread geographically because of the high level of prefabrication used in the scheme, which the consortium said would be the key to its seven-year construction programme. It planned for over 400 basic components of the link to be manufactured at sites spread from St Nazaire in France to the west coast of Scotland, and then transported to the Channel for assembly.

Apart from spreading the benefits of the construction work, this also reduced the amount of construction that would have to be carried out on site in Kent – and EuroRoute argued that the facilities for such a construction programme were already in place in the UK, thanks to the experience of developing the North Sea oil and gas reserves.

The major elements of the scheme were 31 identical 500-metre span, all-steel cable stay bridge sections plus 31 sets of bridge foundations; 84 identical twin-bore steel and concrete immersed tube tunnel road sections, each 250 metres long; 284 steel and concrete sections for the rail tunnel, each 125 metres long; and 36 concrete protective caissons for the islands.

The islands alone were huge projects. They were to be 700 metres long, 500 metres wide and could include shops, hotels, parking for over 1,500 cars and possibly a marina.

EuroRoute estimated that the UK side of the project alone would consume 400,000 tonnes of fabricated steel, 10 million tonnes of concrete aggregates, 6.5 million tonnes of gravel fill, 1.5 million tonnes of cement and nearly a quarter of a million tonnes of reinforcing steel. And on top of all this it would have to build two semi-submersible barges for laying the immersed tube tunnel sections in their dredged trenches, and two precision dredgers to create the trench in which the tunnel sections would sit. EuroRoute estimated that each of these would cost £70 million.

EuroRoute was very much aware that its proposals appeared far out. But it stressed that despite its futuristic image, it was simply harnessing proven technology. "Nothing in our design concepts calls for new or risky construction methods," it said.

But not everyone agreed.

There were other fears too. Some felt that the high cost of EuroRoute would mean the project would have to take a very large share of the cross-Channel freight and passenger market in order to pay its way. This, it was argued, could decimate the ferry industry and leave the promoters with a highly profitable monopoly on which they could charge very high tolls.

These were real concerns. But despite them, EuroRoute maintained its top-level support. So, to the surprise of many – not least the management of the Channel Tunnel Group – Sir Nigel Broackes and his colleagues remained very much in the hunt.

Despite EuroRoute's persistence, to a great many engineers and bankers the race had seemed a foregone conclusion for some time. Given the risks inherent in any construction project – let alone one this big – the simplicity of the rail link scheme and its far lower costs meant it was bound to win, the argument went.

The trouble was that, sensible though the rail tunnels might have been, they seemed a touch, well, dull. And, in any case, this was a project firmly in the political arena, so common sense was not the only criterion.

The Channel Tunnel Group's scheme was born out of the abandoned 1975 project. The consortium that had won that contract was Cross Channel Contractors. It had four members, Taylor Woodrow, Balfour Beatty and Nuttall from the UK and Guy Atkinson from the USA. When the project was cancelled Guy Atkinson left, but the three British firms stayed together with the intention of bidding for work on any future scheme.

Other firms now began putting together their own plans as well, most loosely based on the 1975 scheme. Tarmac and merchant bank Robert Fleming had proposed their twin-bore rail tunnel to be built in three phases.

The European Channel Tunnel Group, whose members included

Costain, German construction firm Holzmann and French builder Spie Batignolles, had put forward five tunnel schemes, four of which were rail-only. Wimpey too were in on the act with Dutch giant Royal Volker Stevin.

By 1984, these disparate groups had gelled into one, the Channel Tunnel Group, with five members, Costain, Taylor Woodrow, Balfour Beatty, Wimpey and Tarmac. What they were planning was in essence very similar to what Tarmac had proposed – two running tunnels for the trains, one in each direction, and a central service tunnel which would also act as the emergency exit – except that they wanted to build it all in one go.

The link would carry through trains between London, Paris and Brussels and would also offer a roll-on/roll-off shuttle train service for cars and commercial vehicles between the terminals at Cheriton near Folkestone and Coquelles just outside Calais.

Following the 1984 banking report, the CTG project was the clear favourite. But although it may have looked a lot simpler than the EuroRoute, it was still a vast undertaking. It comprised three parallel tunnels each 50 km long, two single-track rail tunnels 7.3 metres in diameter and a central service tunnel 4.5 metres in diameter (these diameters were later changed to 7.6 and 4.8 metres respectively). The tunnels would be linked every 375 metres by cross-passages which would both ventilate the main running tunnels and provide emergency access and escape routes.

In addition, there were to be two huge undersea "crossover caverns" – so big some of the engineers nicknamed them "cathedrals" – which would allow trains to cross from one tunnel to another, permitting sections of the track to be closed off for maintenance or in case of accidents.

Tunnelling was to begin from beneath Shakespeare Cliff outside Folkestone and from the bottom of a huge specially sunk shaft outside Sangatte on the French coast. At 75 metres deep and 55 metres wide, the Sangatte shaft was big enough to contain the Arc de Triomphe. From these two coastal sites, the British and French tunnelling teams could work in two directions at the same time, out to sea and back inland towards the terminals. As one tunneller later observed: "It is really a very simple job. It is just very, very big." (It was the length of the tunnel that would later present the project managers with some of their biggest problems. On most long tunnels, materials and equipment can be supplied to the workforce at the face by means of access shafts along the route. With this tunnel undersea for 37 km, that was clearly out of the question, and this meant that logistics and organisation would be crucial to the success of the construction work.)

Whether tunnelling will go well or not is notoriously difficult to predict, even in ground as well surveyed as that between Dover and Calais. But CTG's engineering experts were highly optimistic that the ground would prove excellent. Almost all of the tunnelling could take place in a stratum of chalk marl – an impervious rock especially suitable for tunnelling – around 40 metres below the sea bed, and the tunnels would usually be lined with concrete segments, although cast iron segments would be used in patches of wet ground.

At this stage it looked as though the only problems would occur just off the French coast, where the line of the tunnel would have to rise above the chalk marl and the ground becomes fissured and very wet. This section would require grouting – the injection of a mixture of cement and clay into the fissures – before the tunnel could be bored. But with this known well in advance, the necessary action could be taken and the tunnel boring machines designed to cope.

When complete, the tunnels would carry the roll-on/roll-off shuttle and new high-speed international trains that both British Rail and French railways SNCF believed could take a lot of business from the airlines on the routes between London and Paris and Brussels. The trains, after all, take their passengers from city centre to city centre, rather than leaving them at a distant airport. And SNCF had been keen to tie London into its TGV network for some time.

Shuttle passengers would drive to the Channel Tunnel terminal and load their cars onto specially designed double-deck wagons.

Passengers would have to stay with their vehicles but refreshments and toilets would be available and there would be television screens displaying traffic information. Special lorry shuttles were designed for heavy goods vehicles. Drivers would drive their own vehicles on and off the shuttles, there would be no need to book in advance and the journey time should take half an hour.

Critics asked whether there would be much difference between catching the train and catching the existing ferries, but CTG's backers stressed that it would be much faster and tried to win over some of the pro-drive-through supporters by referring to the shuttle service as a "rolling road". It did not catch on. But although the rail link would create fewer jobs than its more ambitious rival, it would directly employ 4,000 construction workers at peak, and CTG claimed that the total number of jobs created on both side of the Channel during construction would exceed 40,000. (In fact, the tunnel employed 15,000 people at peak construction and created an estimated 70,000 man-years of employment in the UK alone.)

The terminals were huge construction projects in their own right, with the French one taking up as much space as Heathrow Airport.

In all, the project would use up over 70,000 tonnes of structural steel, almost 150,000 tonnes of reinforcement steel, almost two and a quarter million cubic metres of concrete and over six and a half million tonnes of aggregate. And in addition, CTG would need to buy 11 giant tunnel boring machines for the 12 tunnel faces (one could be used twice). And the cost of all this? Just over £2.3 billion at 1985 prices, rising to £4.7 billion when interest payments and inflation are taken into account – a huge amount but half that needed by EuroRoute.

In many people's minds – excluding certain key politicians and large sections of the media – the CTG scheme was the only real possibility if a fixed link was to be built this century. And the events of early 1985 only served to make it a stronger favourite still.

The first news of note was the appointment of Sir Nicholas Henderson as CTG chairman. As a director of Tarmac he had been involved in the project for some time, and had been on the board of CTG from early on. But his move into the chair was astute.

Sir Nicholas was a former British ambassador in Washington and Paris, understood how the French government and civil service worked and had a level of political contacts that was second to none, giving CTG access at the highest levels. His City of London contacts would also prove useful, as he was also a director of both the merchant bank Hambros and the reinsurance group Mercantile and General. Making him chairman gave the project an image and status that, no matter how hard they tried, the heads of the five British construction firms who founded CTG simply could not have matched.

With Sir Nicholas in place, the key issue now facing CTG if the consortium was to launch a successful bid for the mandate was how to bring on board the necessary financial expertise and how to make it a truly Anglo-French project.

As in the UK, major financial and construction firms in France had been looking long and hard at the project and trying to decide whether or not to commit themselves to it. It was a great risk. The rewards were potentially enormous, but so were the potential losses. Even to put together a serious bid for the mandate would cost many tens of millions of pounds.

By the middle of February 1985, within days of Sir Nicholas's appointment at CTG, three of France's biggest construction firms, Bouygues, Dumez and Spie Batignolles, had signed a cooperation agreement meaning that they would their pool their resources to help build a Channel link. Two other top contractors were understood to be on the verge of joining the group, SGE and SAE.

But more significant were the names of the three banks who had also expressed an interest in becoming involved – Crédit Lyonnais, Banque Indosuez and Banque Nationale de Paris. These three were members – along with the Midland and National Westminster Banks in the UK – of the Franco-British Channel Link Financing Group whose 1984 report had done so much to boost the fortunes of CTG's rail link and damage those of EuroRoute.

Just ten days later, National Westminster Bank said it was joining CTG as the group's banker and main financial adviser.

The message was clear. Those financial experts who had looked in detail at the prospects for a fixed link and the schemes on offer were plumping firmly for a rail tunnel, and they were prepared to back their judgement with hard cash.

In May 1985, the French consortium – later to be called France-Manche – underlined the seriousness of its intentions by itself announcing a high-profile president. He was leading industrialist Jean-Paul Parayre, chief executive of Dumez and a former president of car manufacturer Peugeot. But although there was great support in France for the CTG plans, no deal had yet been signed creating the necessary Anglo-French consortium.

According to Sir Nicholas Henderson, who was leading the British attempts to reach a deal, there were a number of problems which if they remained unresolved could seriously affect the rail link's future. One was the uncertainty of the number one figure in France's construction industry, Francis Bouygues, who was keen to explore the possibility of building a cross-Channel bridge, something CTG and the British government had already discounted.

In addition, there were disagreements between the French construction firms and the banks, and between those constructors that were nationalised and those that were not. If such divisions persisted, the opportunity that now existed, thanks to the support for the project from Margaret Thatcher and François Mitterrand, would be lost.

They did not. Heads were banged together and by the end of June pressure from the British had succeeded. There now existed a coherent French team involving five major contractors and three top banks.

On July 2 1985 in London an agreement was signed between CTG and France-Manche. At the same time, Midland Bank, the one remaining member of the Franco-British banking group not in either CTG or France-Manche, said that it was negotiating to join.

The CTG rail scheme had been the insiders' favourite at the start of 1985. Now, the original five British construction firms had top French partners, crucial banking support and a leadership in both Britain and France with the political and diplomatic clout to make things happen. On top of this, CTG had been receiving clear indications of support from international banks, in particular those in Japan. In fact, the response from the international banking community was so good that by October 31, the deadline for handing the mandate bids in to the two governments, CTG and France-Manche had received provisional loan commitments of over £4 billion.

Japanese interest meant that a group of 13 of that country's banks said they would provide £1.64 billion, outstripping the £1.33 billion raised from banks in Britain and France. Surely now CTG and France-Manche were in such a strong position that nothing could stop them?

So it seemed until late October, when James Sherwood, whose firm Sea Containers owned Sealink, decided that if you can't beat them, you might as well join them. Speaking just days before the deadline, he said he would be submitting plans to build his own Channel Tunnel. Vehicles would be able to drive through it; it would carry through trains; and it would be cheaper to build than any other scheme so far on the table. All of a sudden, CTG and France-Manche were no longer clear favourites.

4 Sherwood's surprise

"The most important requirement is that whatever link may be chosen it must be capable of being financed without any support from government funds or government guarantees . . ."

Nicholas Ridley MP, Secretary of State for Transport

Sealink boss James Sherwood was an active supporter of Flexilink, the body set up to oppose the construction of a fixed cross-Channel link of any sort. But hints that Sherwood might also become involved in the competition for the mandate to build the link began to surface in September 1985. Little notice was taken of the news at the time, and no detail emerged until the day that all the schemes were submitted to the two governments.

Announcing that he had entered the race, Sherwood still stressed that he would prefer to see all the schemes abandoned, and he added that when his company bought Sealink in July 1984, for £66 million, he did not envisage that bids to build a fixed link would be sought so soon afterwards. But if the politicians were insisting on pressing ahead with the idea, then he wanted to be involved, he said. He warned that if his bid for the mandate failed, many of the 2,500 people employed by Sealink on the short sea routes across the Channel would lose their jobs. Indeed, in such circumstances, he felt that Sealink's ferry business would be in danger of collapse.

The scheme he put forward was called Channel Expressway, and it must have been music to the ears of the politicians keen to see a drive-through project given the go-ahead. This was especially true of politicians in the UK. Sherwood did not, as yet, have any French partners. Indeed, French wariness of Channel Expressway dogged the scheme to the very end.

In many ways, the Sherwood plan was similar to that proposed by CTG. It involved the construction of two large-bore tunnels, in Channel Expressway's case 11.3 metres in diameter, linking a British terminal at Cheriton outside Folkestone to a French one at Fréthun, near Calais. But unlike CTG's, the Channel Expressway tunnels would each be able to carry both road and rail traffic simultaneously. Every 500 metres there would be crossover tunnels linking the two main bores, allowing sections to be closed off in the case of an accident or emergency and the traffic diverted into a contraflow in the other tunnel. Passenger trains would pass through the tunnel about once an hour, and all freight would go through at night.

The report submitted to Nicholas Ridley explained: "Rail tracks will be set so vehicles will not normally engage them. Motorcyclists will be instructed at the toll plazas not to cross the rail tracks in order to avoid wheel slippage on the rails."

It added that all trains would be hauled through the tunnel by Sealink-owned diesel locomotives. "Before a train enters a tunnel all road traffic will be stopped in the terminal area...a special service vehicle will enter the tunnel behind the last road traffic to ensure that the rail track is clear of all obstructions."

The railways were not happy with the idea of sharing the tunnels with road traffic, arguing that the arrangements would be at best cumbersome and at worst potentially dangerous.

Sherwood's rivals also regarded him, at least at first, rather less than seriously. But if anyone thought that Channel Expressway was only making up the numbers, they had to think again when Sherwood took on board the railways' complaints and changed his plans to meet their demands (much to the irritation of his rivals, who were under the impression that such fundamental changes could not be made after the October 31 deadline).

Now he proposed to build two large-bore tunnels for road traffic and two smaller tunnels for rail. The change would add to the cost, but Sherwood still claimed that his whole scheme could be built for £2.5 billion. SNCF and British Rail were pacified. And what had looked like a rank outsider had begun to move to centre stage.

By the October 31 deadline nine schemes in all had been submitted, of which four complied with the governments' guidelines and were studied in detail by teams of experts.

Apart from the EuroRoute, Channel Expressway and the CTG-France Manche, a group called Eurobridge proposed a series of 5 km span suspension bridges, supported by 340-metre high piers. Vehicles would travel on the bridge on two decks, each of six lanes, enclosed in a tube which would offer protection against bad weather. The promoters also said they would build a 6-metre diameter single-bore tunnel for through rail services.

Eurobridge had some reputable backers, including chemicals giant ICI and construction firms John Laing and Brown and Root, but from early on there were doubts about the scheme's viability. It was the most expensive of the plans – the promoters calculated that construction would cost around £5.9 billion – and many experts had reservations about the group's ability to raise such a large sum.

There was concern that the bridge supports would affect navigation in the Channel and that the International Maritime Organisation might object to it, but most of all, there were doubts whether bridge technology had advanced far enough for such a project to be built: the 5 km bridge spans were each well over three times the length of the 1.4 km central span of the Humber Bridge, which remains today the longest in the world.

Eurobridge said that such huge structures would be possible because of the construction materials it intended to use. "Suspension cables will be made of parafil, an ICI/Dupont material six times the strength of steel for the same weight," Eurobridge said in its description of the project. It said that a polyester reinforced cement called Estercrete would provide a deck surface one-quarter the weight of traditional materials.

The vehicle-carrying tube would be made of Superferrolo, "a Shell concrete with a modulus of rupture more than half that of the volume equivalent of mild steel at a third of the weight."

It added: "With its ability to withstand such substances as concentrated sulphuric, hydrochloric and nitric acids and most organic solvents, it is a very effective proof against corrosion and is thus maintenance-free."

It was certainly a project for the future, and it touched a raw nerve with some CTG supporters because it made their rail link look somewhat staid and unexciting. But Eurobridge gained a rather eccentric image – an image exacerbated in the eyes of many journalists when a paper produced at a press conference listing those involved in the project contained a note saying it had been copied to "HRH Prince Alexander of Jugo-Slavia" with no further explanation – and civil servants close to the Channel link debate clearly thought the project was pushing the frontiers of bridge technology too far, too quickly. It did not become a serious contender.

Although none of the other schemes met the governments' published criteria and so were not considered for the mandate, they at least showed that the spirit of invention was not dead in either Britain or France.

According to the French, one scheme, submitted only in Paris by a Mr Van der Plutten, involved blocking the Channel with two tidal power stations, one on each coast, leaving a 6 km wide gap in the middle for shipping to pass through. The idea was that the income from the electricity generated would ensure any fixed link's financial viability.

Speaking in the House of Commons on November 1 1985, after receiving the submissions, Transport Secretary Nicholas Ridley again stressed, in case there should be any doubt, that the project would only go ahead as a purely private sector enterprise. "Today is a very important day in the long saga of whether there should be a fixed link between England and France.

The most important requirement is that whatever link may be chosen, it must be capable of being financed without any support from government funds or government guarantees against commercial or technical risks."

He continued by setting out the future timetable of the project, so simultaneously underlining both governments' commitment to the task. "With our French colleagues we shall conduct our evaluation of these proposals with great care. Our aim is to decide early in the new year which project, if any, is to go ahead, and to announce our decision by mid-January. If that timetable is met, the way will be clear for conclusion of a Treaty with France towards the end of February and the introduction of legislation into Parliament by next Easter, with construction of a link beginning in mid-1987."

There could be little doubt now that the governments were going to do all they could to get a scheme off the ground – bar paying for it. The key remaining question was which scheme?

If any of the various promoters thought that their work was done, and all they now had to do was to sit back and wait for the government's decision, they were sadly mistaken, for if anything the public relations battles became even more fierce in the weeks ahead.

To outsiders, it looked as though Flexilink and their anti-link allies were fighting a losing battle. But they had not given up all hope and believed that they could yet make life very uncomfortable for the British government and the link's supporters.

At Westminster, a full Commons debate on the link was due to be held early in December, and before then the House of Commons Transport Committee was due to publish the findings of its own investigation. At the same time, the Anglo-French team assessing the merits of the various schemes for the two governments was due to complete its work before Christmas, in time to allow ministers in London and Paris to consider its findings. Then Thatcher and Mitterrand would announce the winner at a special ceremony on January 20 1986 in Lille, northern France. Every opportunity to lobby ministers, civil servants and MPs had to be grasped.

If the pressure was being felt anywhere, it was in the offices of EuroRoute. On November 1 1985 there were three serious contenders for the mandate. Within days, it began to look like a two-horse race, at least in London, with James Sherwood displacing Sir Nigel Broackes as the drive-though supporters' favourite. Sherwood also had better relations with the railways because of EuroRoute's plan to open the rail link 18 months after the road link had been completed.

The railways feared that EuroRoute's scheme would mean they would lose a lot of freight business to road hauliers. The plans were later changed so that one single-track tunnel would be opened at the same time as the road link, with a second to be added later according to demand but relations between EuroRoute and the railways remained strained.

Some insiders wondered whether Sherwood's original reason for entering the race for the mandate had been to destroy EuroRoute's chances of winning. There was a simple logic to it: City financial experts had long been arguing that for EuroRoute to justify its high construction costs, it would have to capture a very large percentage of the cross-Channel ferry market, so jeopardising those routes' existence. CTG, on the other hand, had always said it saw itself competing with the ferries in an expanding market.

Such considerations rapidly became irrelevant, however. If Sherwood had originally entered the race as a spoiler, he was much more than that now. By early December he was claiming to be the front runner. On the eve of the Commons debate, Sherwood told the *Daily Telegraph*: "I believe we are in the lead. I am very pleased." He added that a key factor in his success to date was the good relations he had developed with the railways, particularly SNCF in France. "We are closer to the railways than the other bidders," he said.

CTG was faced with a new, and serious challenge: having led the race from the start, it was now in danger of seeing the prize snatched from its grasp at the last.

EuroRoute meanwhile was slipping further behind. Its financial backing looked less certain than CTG's, relations with the railways were rocky, and concern was growing about its technical viability.

A report in the *New Civil Engineer* magazine – which is owned by the Institution of Civil Engineers – said that some engineering experts were strongly critical of the scheme, and added that what it called "technical and financial" problems had all but ended EuroRoute's chances. The engineers pointed out that some elements of the proposed technology had never been tried in a hostile environment.

Bob Sellier, head of EuroRoute's construction arm, condemned the *New Civil Engineer* report as "absolute rubbish", but it was clear that in Westminster, if not in Paris, EuroRoute was struggling to stay in contention.

On December 10 1985, the Commons debate on the principle of a Channel fixed link passed without a hitch.

Six backbench Tory MPs voted against the government – Jonathan Aitken, Peter Rees, Roger Moate, Patrick Cormack, John Stradling Thomas and Teddy Taylor – but it still won the vote with a majority of 96.

Nicholas Ridley also confirmed that no public enquiry into the scheme would be held – something that Labour members had been pressing for – as he feared the delays and uncertainties caused by such a move would scare off private investors. Instead, after the Treaty with France had been signed, Parliament would be presented with a hybrid bill – that is, a public bill that affects private rights – and opposition could be voiced either on the floor of both houses of Parliament or by giving evidence to the various Parliamentary committees that would study the bill. Critics in Parliament and outside said this was an outrageous infringement of the democratic rights of the scheme's opponents, but there was little they could do.

Four days before, CTG's rail link had received a boost when the Commons Transport Committee voted narrowly in its favour because it was less of a financial and technical risk than any of the other schemes and because it would have a smaller impact on the environment. But the vote was close – it hung on the casting vote of the chairman, Gordon Bagier – and so its conclusions would carry little political clout.

Within a week, Sir Nicholas Henderson and his CTG colleagues would have been excused for putting the champagne on ice. Stories began to leak into the press that the French experts studying the various schemes had come down firmly in favour of the CTG rail tunnels. If the British assessors drew the same conclusion – as had just about every team of experts that had studied the options in detail to date – then surely they had seen off the challenge from James Sherwood.

Supporters of CTG found it difficult to understand why the British government was taking Channel Expressway so seriously. Sir Nicholas Henderson later wrote: "I attempted to discover whether and why Sherwood really had the inside track he was claiming. It seemed odd to me that he should apparently be so favoured. He was a ferry operator who initially had been against any idea of a fixed link; indeed, he was a member of Flexilink, an organisation whose purpose was to oppose the creation of any fixed link. He had only put in an offer for a fixed link at a very late stage."[1]

Certainly, there were very real question marks over Channel Expressway. One was cost. The magazine *New Civil Engineer* quoted an unnamed, independent tunnelling engineer who had examined all the schemes and decided that Channel Expressway's costs were about 50 per cent too low. The Transport Committee report had touched on the same point, saying that the costs of Sherwood's scheme "presented a greater challenge to credulity" than those presented by CTG and EuroRoute.

CTG managing director Michael Gordon stressed the point in a letter to the *Financial Times*: "It is our view that the cost being quoted for the Channel Expressway scheme is wildly underestimated. A road tunnel involves a much larger bore than a rail tunnel and this presents serious engineering problems requiring costly solutions." He went on: "Construction times and costs relating to this scheme have not, we believe, been fully investigated. Moreover, the costs allocated for the construction of a twin rail tunnel appear to bear no relation to all the available evidence in this field."

There were other worries, too, such as the effect on drivers of going through such a long tunnel. James Sherwood had met these concerns head-on in typically vigorous fashion in his original submission document. "Opponents of drive-through tunnels, perhaps better described as the rail lobby, have argued that drivers will become fatigued or mesmerised by driving 30 minutes through a long straight tunnel," he wrote. "There is not a shred of evidence to support the allegations of driver fatigue or uncontrolled operation of motor vehicles in excess of that incurred on motorways by virtue of driving through a 50 km properly designed tunnel." And he added: "While the Channel Tunnel would be the longest drive-through tunnel in the world, there is no evidence of such a problem in existing tunnels such as the Mont Blanc (11.6 km) and St Gothard (16.3 km)."

Others were not convinced. CTG commissioned research from consultants Environmental Resources which said that even in the tunnels through the Alps, some drivers have difficulty keeping to the lanes. And it referred to research carried out in Japan which revealed a high rate of accidents in tunnel areas, especially among older people, and which called for urgent investigation of the causes of such accidents and what to do about them. MPs on the Transport Committee conceded that driver psychology was a real issue in a road tunnel as long as that planned by Channel Expressway. But there was one doubt which dominated all others. And that was how the road tunnels would be ventilated.

Sherwood acknowledged the concerns: "Ventilation is the most serious technical challenge of a drive-through tunnel." But it was a problem he believed he had solved. Ventilation systems in road tunnels have to deal with carbon and soot deposits as well as dangerous gases such as carbon monoxide. Channel Expressway intended to counter the problem of the carbon and soot particles by installing electrostatic precipitators in by-pass tunnels. Powerful fans would draw the air into the by-pass tunnels and through the electrostatic precipitators, which, experts said would remove 85 per cent of the smoke particles. The level of carbon monoxide would remain unaltered and would have to be diluted with fresh air. But supporters of Channel Expressway said that such a system would reduce the number of mid-Channel ventilation shafts needed to two, which could both be located outside the main shipping lanes. "This technology has been developed in Japan and proven to be safe and successful," Sherwood claimed. Up to a point, said his opponents.

CTG claimed it had carried out studies into the possibility of building a drive-through tunnel earlier in 1985 and concluded that such a project was not practical or financially feasible.

In the same letter to the *Financial Times* as that quoted above, Michael Gordon said that the carbon monoxide from exhaust fumes could not be reduced "to tolerable levels" using known technology. Ken Groves of EuroRoute added: "To date the technology has only been used to ventilate three tunnels, the longest of which is just 11 km. We cannot see the feasibility of adopting it into the Channel Expressway – a tunnel of three times that length."

Questions were also asked about whether the ventilation system could cope in the event of an accident. Would the electrostatic precipitators be overloaded by a localised build-up of smoke?

The engineers and designers responsible for Channel Expressway hit back. The system was not as revolutionary as it looked, they argued. The plant required for even the longest stretch of tunnel was already available and required no further development. What is more, the job of ventilating the tunnel would become easier still as exhaust emission standards improve in line with new legislation in the years ahead.

And as for the prospect of accidents, they would be kept to an absolute minimum because traffic in the tunnels would be kept at a very low density.

The debate raged on.

On top of all this, James Sherwood was still lacking the level of French support enjoyed by his rivals. Admittedly, French construction firm Screg had thrown in its lot with Sherwood, who had been working hard to develop his political and financial contacts on that side of the Channel. But many observers argued that no matter what the British government decided, Sherwood would find it impossible to convince the French to back a scheme that would create far fewer jobs in France than its rivals. Therefore, they argued, Channel Expressway was ultimately doomed.

Yet despite all the criticisms and the doubts voiced by experts, it remained a serious challenger. Indeed, in some quarters, and especially in Britain, support for Sherwood seemed if anything to be growing.

In an editorial, the *Financial Times* said that "it seems unwise to settle for a scheme that does not include a road link". It added: "The cheapest and most practical on offer appears to be that of Channel Expressway." And there was no doubt that the British public liked Sherwood's plan. A MORI opinion poll came down firmly in favour of both a drive-through scheme and Channel Expressway's version of it.

The year ended in confusion. Logic dictated that CTG and its rail link should have had the mandate all but in the bag. The majority of independent construction and financial experts seemed to be backing it. But with both Mrs Thatcher and President Mitterrand both still hanging onto their ambitions for a drive-through link, anything could happen. The competitors had spent millions on design work, publicity and lobbying. Yet with 20 days left, the race was still wide open.

[1] From *Channels and Tunnels*, by Sir Nicholas Henderson, published by Weidenfeld and Nicolson, 1987

5 Concessions and treaties

"You can't say this will be an absolute bonanza for UK industry. If the price is right it will be, if the price is wrong it won't be."

Richard Hannah, UBS Phillips and Drew

The behind-the-scenes manoeuvrings now started to reach new heights. Rumours abounded. Perhaps all three groups would merge and then split the work between them? Perhaps the rail-only Channel Tunnel Group would take as a partner one of the drive-through schemes? More certain was the split between Britain and France over James Sherwood and his Channel Expressway. A French government official said Paris was very worried by the lack of "serious industrial partners" in the scheme.

There was a definite smell of fudge in the air. On January 10 1986, just ten days before the winner was due to be announced, the *Financial Times* reported: "It seems likely that the announcement to be made in Lille, northern France, when Mrs Thatcher visits on January 20, will fall short of approving one single project."

CTG supporters were exasperated. By now it had become clear that the British as well as the French experts on the Franco-British assessment group – in fact the group was divided into eight sub-committees covering the environment; law and finance; construction; frontier controls; security; employment; maritime issues; and traffic – had recommended to ministers that it should choose the CTG all-rail link. This was the group set up by the two governments to look in detail at the pros and cons of all the projects. Surely ministers would have to abide by its recommendations? They would be taking a huge risk if they ignored its findings and then something went wrong.

As far as CTG was concerned it had won all the arguments. Its scheme was less of a technical and financial risk than any of the others. And CTG's backers firmly believed that a drive-through link would be too costly and would push the available technology to its limits and probably beyond. Yet still the politicians on both sides of the Channel wanted to be able to drive across the Channel.

For Sir Nicholas Henderson and his colleagues the only consolation was that one government favoured EuroRoute and the other Channel Expressway. But CTG was now in danger of being runner-up in both France and the UK, albeit to different schemes in each country.

Just before Christmas 1985, there were reports that the French government had suggested a merger between CTG and Channel Expressway. Early in the new year, this was raised repeatedly by Transport Secretary Nicholas Ridley with both James Sherwood and Sir Nicholas Henderson. The idea was firmly rejected by the CTG board, but it did agree to make a crucial concession – it said that if demand and technology permitted, it would itself build a drive-through link in addition to its planned rail-only scheme.

Sir Nicholas set out CTG's position in a letter delivered to Mr Ridley on January 10 1986 and later released to the press. He repeated that CTG did not think a drive-through scheme was currently feasible. He wrote: "When I saw you yesterday you put to me the suggestion that the Channel Tunnel Group/France-Manche should get together with Channel Expressway with a view to submitting a joint proposal for a fixed link to the French and British Governments...The Channel Tunnel Group Board have asked me to notify you in the most categorical terms that they are not prepared to enter into any arrangement of the kind proposed."

He added that CTG had looked in detail at the sort of scheme proposed by Sherwood and had firmly rejected it. "It would be extremely costly and the risks...were such as to make impossible the financing of such a scheme in the private sector." But he then concluded: "However, if the traffic warrants it, if there is demand for a drive-through and if the various problems and uncertainties of a drive-through can be met in a way compatible with market financing, the Channel Tunnel Group and France-Manche will be prepared to develop and implement a drive-through scheme.

"It is not possible to say when this might occur because it depends on developments both in traffic and technology. However, Channel Tunnel Group and France-Manche are prepared to give this undertaking now and consider that account should be taken of it by the governments in considering the Cross Channel Link." It was. Those last two paragraphs allowed the politicians to appease public opinion and say a drive-though link would be built – as soon as it was technically and financially possible. That was to prove crucial.

On Thursday January 16 1986, the Cabinet met in London. The Channel link was on the agenda but no announcement was made. That afternoon,

Nicholas Ridley met his opposite number, French Transport Minister Jean Auroux, in Paris. The meeting was apparently amicable, but again there was no announcement.

By now, almost all the pundits were backing CTG. The British newspapers were convinced a final decision had been taken and that the rail link had won.

That same evening, Sir Nicholas Henderson received a telephone call that left him in high spirits. He had been due to spend the weekend with the Foreign Secretary Sir Geoffrey Howe and his wife Elspeth, but at the last minute Sir Geoffrey rang to cancel.

"He thought it would be awkward if it came out afterwards that I had spent the weekend with him," Sir Nicholas wrote. "In fact the postponement gave me considerable encouragement because I deduced – not that it required a blinding flash of imagination – that if it had been decided by the Cabinet that we were not the winners, then there would be no particular embarrassment to Howe or others if it came out that we had spent the weekend with the Foreign Secretary." [1]

After the brinkmanship and the furious lobbying of early January, it was something of an anticlimax to hear Prime Minister Margaret Thatcher and President Mitterrand announce in Lille that the Channel Tunnel Group and France-Manche had indeed won the mandate.

There was all the pomp and glitter expected of such an occasion.

The bands played – for some reason the repertoire included "It's a long way to Tipperary" – a great deal of champagne was drunk and the food was splendid. The necessary speeches were made. Everyone said it was an event of great significance for the two countries and of course for the contractors, engineers and bankers involved. The joint British-French communiqué published on the day talked of the enthusiasm for the project and its "symbolic nature". It added: "It testifies to the willingness of the two countries to strengthen their economic, political and cultural ties and to demonstrate to future generations an example of imagination and enterprise for peaceful purposes." Heady stuff.

But already the minds of those involved were turning to the work ahead. Jamborees such as this were all well and good, but tunnelling was due to start in 18 months' time. Before then legislation had to be passed, money raised and designs finalised. And there was also the thorny matter of agreeing the construction contract with the five British and five French contractors who were going to build it. What that contract included and excluded was an issue that would dominate the project for years to come (indeed it looks set to exercise the minds of some of the best construction lawyers for years yet).

The weeks after the award of the mandate saw a flurry of activity. The

Channel Tunnel's opponents gained a new lease of life. The French government announced a massive infrastructure investment package in northern France to support the Channel Tunnel, and the Channel Tunnel Group began a major restructuring and recruitment exercise in preparation for the daunting financial and civil engineering tasks ahead.

In some ways, the project's opponents now had an easier job. At least they had a specific target. Environmental groups clubbed together and repeated their call for a public enquiry. They were especially critical of the siting of the shuttle terminal in Kent, and felt it should be located closer to London, so reducing the strain on the county's roads.

A clear majority of MPs were in favour of the project in principle, yet many on both sides of the House of Commons were unhappy at the level of consultation, as were the inhabitants of Kent. Others feared the tunnel would exacerbate the UK's north-south divide, bringing jobs and investment to an already prosperous southeast, with the north cut off from the boom because of poor transport links.

These opposition voices were joined by those whose livelihoods would be affected by the tunnel: the ferry operators and crews and the Kent businesses that relied on them – and of course the losing bidders for the mandate.

Sealink and Channel Expressway chairman James Sherwood said the decision would mean savage cutbacks in the level of ferry services with little benefit for the travelling public. "It is certainly a sad day for the motorist who will be offered an option only marginally better than the ferries," he said of CTG's scheme. And he served notice that the battles to stop the project would continue for a long while yet. Legislation had to be passed by both the French and British legislatures before construction could start. It could expect a difficult ride. "The polls of the general public in both Britain and France show that 75 per cent of those that want a fixed link want a drive-through solution. Governments which ignore the wishes of the public do so at their peril." And he added: "If the socialists are turned out of office in France next March we have received assurances from the centre and right political leaders that the necessary legislation to allow the all-rail link to be built will never be passed."

There were other signs, albeit localised, of growing disquiet in France. The Mayor of Calais, Jean-Jacques Barthe, said he feared that the planned port extension programme would be scrapped because of the tunnel and that thousands of jobs would be lost in the town, which already had an unemployment level of 17 per cent. His concerns were echoed by the local Chamber of Commerce and the trade unions.

In the UK, the anti-Channel Tunnel group Flexilink accused the two governments of pressing ahead with a link without taking full account of the consequences. "In the long run any such project will destroy thousands of jobs, cost more than the ferries and provide travel scarcely quicker than currently available," it argued in a statement issued immediately after the Lille festivities. And it said that it would be the Channel Tunnel not the ferry companies that would struggle to survive. "The existing cross-Channel services can compete effectively and decisively with any form of fixed link and destroy it financially."

The unions too were adding their weight to the protests. Ron Todd, the then general secretary of the Transport and General Workers Union, said the question that had not been asked was whether the Channel Tunnel was worth the amount of money that would be spent on it. "There is a question of social priorities. A better health service, more housing and modern schools, more care and facilities for the old and handicapped – these are surely crying needs and better ways of spending money. Our sewage system is collapsing and our roads need more money spent on maintenance," he said. "These are the infrastructure projects that should have priority – not grandiose schemes whose purpose it seems to me is to provide a monument to political egos on both sides of the Channel. That is why we still need a public enquiry...into the whole issue of whether a Channel link is needed or should be a priority for Britain."

He looked back to the days when one of his predecessors, Jack Jones, "was very effective in behind-the-scenes moves which led the Labour government to cancel the Channel Tunnel in 1975".

But times had changed. Union power was on the wane; the British government had a strong-willed leader with a huge majority in the House of Commons and she was backing the project; and in France, although there were outspoken opponents, the project simply did not arouse the passions that it did in Kent. Indeed, there was a rumour that an opinion poll in the south of France found that a sizeable minority of respondents thought that a Channel Tunnel had already been built. Apart from the occasional headline, it was still hard to see what the Channel Tunnel's opponents could achieve.

What is more, for every voice against the project, there were many more in favour. Construction plant and materials firms on both side of the Channel were licking their lips at the prospect of £1.4 billion worth of orders which CTG expected to split evenly between Britain and France. The shopping list was very long and very expensive. On the UK side, CTG expected to spend £150 million on construction equipment such as concreting plant, conveyor systems, cranes, earth-moving plant, pumps and the tunnel boring machines; £240 million on construction materials such as aggregates, reinforcement, structural steel, shuttering timber and tunnel linings; £150 million on the mechanical and electrical systems; and a further £160 million on signalling, overhead line equipment, track and the shuttle trains for the tunnel railway.

City analysts, including Channel Tunnel expert Richard Hannah of

UBS Phillips and Drew, warned that jobs and orders would not automatically stay in the home countries. The pressure would be on CTG to find sources at the right price and quality, and as this was a purely private sector project that would take precedence over the nationality of the supplier. "You can't say this will be an absolute bonanza for UK industry," Mr Hannah said at the time. "If the price is right it will be; if the price is wrong it won't be."

British CTG managers said that their basic procurement policy was to buy British whenever possible, and they believed that most materials and equipment could be sourced at home. But with all supply contracts going out to tender, that looked somewhat optimistic. One area that typified this apparent conflict of interest was the supply of the six tunnel boring machines (TBMs) required for the British half of the contract.

Although tunnelling technology was available in Britain, there were many overseas manufacturers – in the USA and Japan especially – more than able to supply the required equipment.

Construction News reported just three days after the award of the mandate that foreign firms might import their TBM technology into Britain, where the machines might be manufactured under licence. "To this end, Markham, based in Chesterfield and Robbins from Washington, USA, have entered a manufacturing arrangement under which Robbins' full-face tunnel boring machines will be built in this country. The two firms will consequently be presenting a united tender for the supply of TBMs for the contract, which must make them serious contenders for the award." So it proved. The Robbins-Markham joint venture supplied two of the six British TBMs, with the balance coming from James Howden of Glasgow.

Despite the threat of strong overseas competition, British industry was, of course, delighted at the boost to business the Channel Tunnel would bring. But industry leaders were already concerned about how passengers were to get to and from the tunnel and whether the government was doing enough to ensure that Britain took full advantage of it.

Dennis Cooper, chairman of the CBI in Kent, fired an early shot over the Department of Transport's bows. He told ministers that major road improvements should be carried out before the tunnel construction got into top gear.

Better links to the M25 were crucial. In addition, "roads should be improved northwards to the Thanet area and northwest to the M2 link" and the government should commit itself quickly to these and other measures. There was also the small matter of the rail links to the tunnel for the through trains. Many of Kent's existing commuter lines were already on their last legs – how could they be expected to cope?

Mr Cooper's voice was one amongst a growing chorus and the anxiety felt in the UK that not enough was being done to prepare the country for

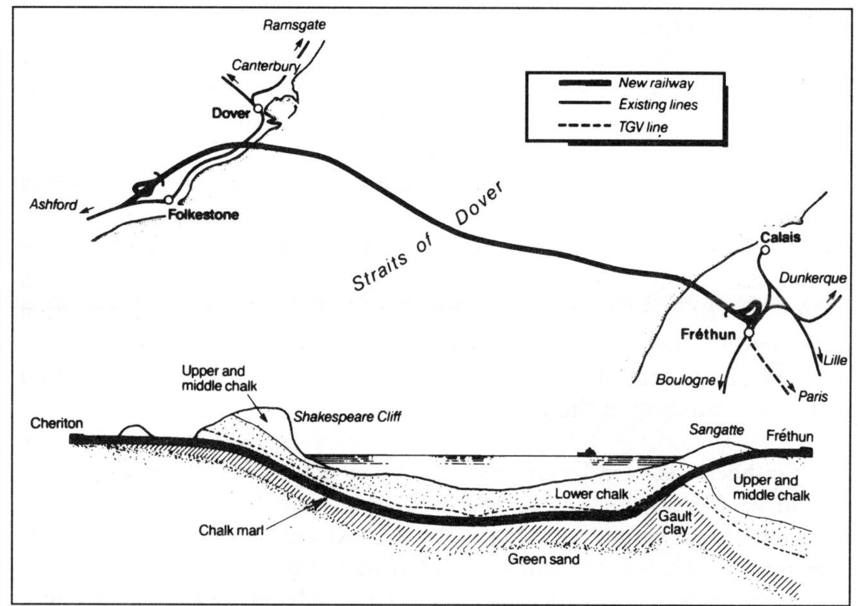

The route of the winning scheme.

the tunnel was exacerbated by the huge package of measures announced by the French government within days of the Lille celebrations.

The French railway SNCF was already planning to extend its TGV high-speed rail network north, linking Paris to Lille, Amsterdam and Cologne, with a spur through the tunnel to London if agreement could be reached with British Rail.

Now the French government announced details of "Le Plan Transmanche", a six-billion franc package of transport improvements in northern France coupled with support for the Channel ports. The package included extending the A26 motorway into Calais; building a motorway along the coast linking the Belgian frontier with the mouth of the Seine; improving port facilities at Calais, Boulogne, Dieppe and Dunkerque; modernising the local rail network between Calais, Amiens and Boulogne; and creating a coastal development fund to help boost tourism.

In stark contrast, the British government's response looked piecemeal and penny-pinching. Admittedly, road improvements in Kent were on the drawing board, most crucially the extension of the M20 motorway. And British Rail later confirmed it would be spending around £400 million on new trains, line improvements, maintenance facilities and international stations at Waterloo in London and at Ashford in Kent (although work has yet to start at Ashford, almost eight years later).

But even though the British economy was looking stronger by the day, British ministers seemed to lack the enthusiasm or imagination of their

French counterparts. Within days of the mandate award, Transport Secretary Nicholas Ridley had ruled out any massive improvement in the rail system from Folkestone to London. The improvement of track standards to take high-speed trains "would cause major disruption and inflict immense damage on the environment", he said.

In 1994, "Le Plan Transmanche" and the TGV Nord are just about complete. Meanwhile, British ministers are still talking about how a high-speed rail line linking London and the Channel Tunnel, a distance of only 70-odd miles, might be financed and built.

Why has Britain been unable to plan and build such an important piece of transport infrastructure, even when money was readily available during the late-1980s economic boom? This question was to be asked with increasing irritation in the years ahead.

For Sir Nicholas Henderson and his colleagues at the Channel Tunnel Group, however, there were more pressing matters to attend to than whether Britain would be able to take best advantage of the tunnel when it opened. There was the small matter of building it.

On the award of the mandate, CTG immediately set about a major reorganisation. It started by splitting itself into two. In future, there would be an owner, Eurotunnel, and a contractor, Transmanche Link. Eurotunnel would have the job of raising the money, holding the concession agreement from the two governments and running the tunnel for a profit after it had been completed.

Eurotunnel would also let a design and build contract for the tunnel to Transmanche Link, a consortium of the five British and five French contractors that originally founded the Channel Tunnel Group and France-Manche. The contract involved not just building the tunnels but designing and installing a complete railway system. The idea was that all that Eurotunnel would have to do would be to take delivery of the completed project in just over seven years' time and test it ready for the opening a few months later.

In February 1986, shortly before the Channel Tunnel Treaty was signed by Britain and France in Canterbury, Sir Nicholas Henderson said he would be resigning soon. He had always said he wanted to leave when the mandate had been awarded, so he had in fact stayed on slightly longer than intended. But the departure of one of those whose efforts had been so influential in CTG's victory underlined the scale of the changes under way.

The close relationships nurtured by Sir Nicholas between contractors and bankers were necessary for a successful bid but were not necessarily appropriate with the team split into the adversarial roles of client and builder. In years to come, many involved in the project would look back at this time and ask if the split had been necessary for the project to proceed. Perhaps work would have progressed far more smoothly if the client and contractor had not been created as separate entities.

But they were, and for the moment at least, everything seemed to be running smoothly. Banking sources said the details of the construction contract were close to being agreed and should be finalised by mid-February. Negotiations between the Channel Tunnel Group, France-Manche and the two governments over the fine print of the concession agreement – the formal agreement which gives the mandate winners the right to build and run the link, and which would determine such issues as how long the mandate should last for – were taking longer than hoped, but they were making progress.

On February 12 1986 in Canterbury, the Lille announcement was confirmed by the signing of a treaty between Britain and France in a ceremony with just as much pomp as its counterpart a few weeks before. *Construction News* reported at the time that the publicity "would have done justice to an accord to abolish nuclear weapons...the sedate city of Canterbury must have wondered what had hit it. The electronic media brought so much hardware with them that the ancient hall assigned for the press's use bore a passing resemblance to mission control."

The project's opponents too made their presence felt, and showed that they were not going to give in lightly. A few hundred anti-link demonstrators had mustered outside Canterbury Cathedral as the ministerial motorcades swept in, chanting slogans such as "Maggie, Maggie, Maggie, Out, Out, Out" and, in honour of President Mitterrand, "Froggy, Froggy, Froggy, Out, Out, Out."

One scored a direct hit on the President's Rolls Royce with an egg. They did not disturb the ceremony itself.

Under the French constitution, the President does not sign treaties, so that duty fell to the French foreign minister Roland Dumas and Sir Geoffrey Howe, his opposite number.

But Thatcher and Mitterand did not stay out of the limelight for long. "Deprived of centre stage, both leaders appended their signatures to everything they could – the visitors' books of the King's School, the deanery and the Canterbury Cathedral itself," *Construction News* commented. The two also made short speeches which, with hindsight, provide a small insight into their different attitudes towards Europe.

Mrs Thatcher primarily focused on the concerns of the people of Kent about the effects of the tunnel on their day-to-day lives.

"We shall ensure that everything possible is done to mitigate the environmental consequences [of the link]," she said. "We are also determined to do what is necessary to improve the local road network so that there can be rapid and uncongested access to the link." She added: "I would urge the people of Kent...to treat the link as an opportunity that will bring long-term and lasting benefits – tourism industry and jobs – to the region and to the United Kingdom as a whole." But President Mitterrand stressed the wider implications of the project, seeing it within the context of growing European cooperation and unity: "It is in fact a part of the European infrastructure that we are about to build."

Just over a month later, on March 14, the tunnel took another important step forward. The Channel Tunnel Group and France-Manche and the British and French Transport Ministers, Nicholas Ridley and Jean Auroux, finally signed the concession agreement.

There was never any doubt that it would be signed, but the document was important nonetheless.

In essence, the concession laid down the reciprocal commitments of the two states and the concessionaires. The agreement would come into force in the summer of 1987 after the British legislation had been passed by Parliament and had received Royal Assent. It confirmed that the project would be carried out without government funds or guarantees "of a commercial or financial nature"; granted the promoters a 55-year concession period; and it said that preparatory work for the tunnel should be finished in three years, the service tunnel breakthrough achieved within seven years and the whole project completed by 1997.

The winning promoter had been announced, the treaty between Britain and France was signed, and the concession details agreed between the promoter and the two governments. But as yet there was no construction contract, no money, and the necessary legislation still had to be passed by MPs in Britain and France. There was an awful long way to go yet before the first sod could be turned.

[1] From *Channels and Tunnels*, by Sir Nicholas Henderson, published by Weidenfeld and Nicolson, 1987

6 Money and politics

"Eurotunnel's sums just do not add up."

Jonathan Sloggett, Chairman, Flexilink

Just days before the concession agreement was signed, the Channel Tunnel Group announced that Sir Nicholas Henderson's replacement as chairman would be Lord Pennock. It was a choice designed to please the banks and the City of London. He was a director of the merchant bank Morgan Grenfell and of the Standard Chartered Bank, and was also president of the European Employers' organisation UNICE. Previously he had been deputy chairman of ICI, president of the Confederation of British Industry and, from 1980 to 1985, chairman of BICC, whose construction subsidiary Balfour Beatty was one of the British founder members of the Channel Tunnel Group. Lord Pennock was a heavyweight experienced industrialist, but even he must have wondered what on earth he had let himself in for. Within weeks of taking up the post, he was beset by problems.

Opposition to the project was becoming more vocal. And it was beginning to make itself felt in Parliament, something that had seemed unthinkable even a few weeks earlier. The progress of the hybrid Channel Tunnel Bill through the House of Commons was now looking anything but smooth. Some MPs were predicting a six-month delay and that would mean putting back the start of construction.

Talks with the contractors about the contract had not gone as well as had been hoped. The complexities of the project meant that the contract was still some way from being signed. And all the above meant the bankers and the City of London were getting twitchy.

When the concession agreement was signed, CTG and France-Manche began the process of merging, creating Eurotunnel (it formally came into being on April 28 1986) with Lord Pennock as chairman. A French co-chairman would be appointed in the autumn. Eurotunnel would be the owner of the Channel Tunnel and hence the holder of the concession.

At the same time the contractors split apart – although they retained a stake in Eurotunnel – creating Transmanche Link, an Anglo-French joint venture which was to have the job of building and commissioning the scheme. John Reeve of Costain and Philippe Montagner of Bouygues became TML's joint directeurs generals.

On April 18 1986, Transport Secretary Nicholas Ridley published the Channel Tunnel Bill, which would allow the government to ratify the Treaty with France and which would give Eurotunnel the authority to construct and operate the project.

Suddenly, the Channel Tunnel's luck seemed to run out, with the news that the legislation could be delayed by weeks, or even months. The problem was that the Bill was in breach of standing orders because it had missed its official deadline and hence could be thrown out unless it was granted special dispensation allowing it to be treated as urgent business. The Department of Transport had also incurred the wrath of Deputy Speaker Harold Walker, because it had circulated an unofficial timetable for the Bill's passage through Parliament which critics said stifled debate. The result was that the House of Commons Standing Orders Committee was convened, chaired by Harold Walker, to decide whether the Bill should be allowed to proceed to its second reading or whether it should be rejected. If the committee decided to do the latter, a new Bill would have to be drawn up and re-presented to Parliament, probably the following November.

The snag irritated the new Eurotunnel chairman no end. Lord Pennock accused MPs of "mucking about", and said delays of this sort would deter international financiers from investing in the project. The Standing Orders Committee met on May 20 1986. It heard opponents of the tunnel, such as the Dover Harbour Board and Jonathan Aitken, the Conservative MP for Thanet South, argue for delay, and the Department of Transport calling for the Bill to be allowed to continue. When it came to the vote, the committee was split, 5-5. And to make matters more confusing still, chairman Harold Walker declined to use his casting vote. Instead, he referred the matter to the whole House.

For Eurotunnel, though, this was better than it looked. For with an overwhelming majority of MPs from all sides in favour of the Channel Tunnel, success at that later stage on the floor of the House of Commons was virtually guaranteed. A six-month delay had successfully been averted.

While this minor Parliamentary drama had been unfolding, talks between the contractors Transmanche Link and Eurotunnel on the contract had been making progress. There was pressure to come to a deal quickly because raising money would be nearly impossible if Eurotunnel could not tell potential investors in some detail what the tunnel would cost to build.

The contract was actually awarded to TML on May 15 1986. It entailed designing, building and commissioning the entire project in just seven years, ready for opening in May 1993. The contractors themselves described the task as "formidable", as it included not just the building works but an entire transportation system.

In a briefing for journalists held at the time, TML said: "We are contracted to construct not just the rail and service tunnels under the Channel but also the passenger, rail and freight terminals on either side. In addition, we are responsible for the creation of the special shuttles, the track itself, the advanced signalling and control systems – and a wide spectrum of other services needed to allow high-speed passenger and freight trains to travel between Britain and France."

This was far more than a civil engineering job and would push the expertise and organisation of the TML member firms to the limit. Some of the technology involved, for example the signalling system for the high-speed trains, was the most advanced of its type in the world.

Although scope of the contract was known, the detail was still being discussed and nothing had as yet been signed. Even so, preparatory work was already under way. The engineering designs were being drawn up and preliminary surveys carried out in the Channel. TML was drawing up plans to set up its own factory on the Isle of Grain in northern Kent to manufacture the concrete tunnel linings. In France, work was beginning on the huge shaft on the coast outside Calais, from the bottom of which the tunnelling machines would start heading both out to sea and back inland towards the terminal.

By June 1986, the Channel Tunnel Bill had received its second reading (309 votes for, 44 against). Despite this, Thanet South MP Jonathan Aitken gave notice that he and other anti-tunnel protesters would not be easily ignored.

The Bill now went before a House of Commons Select Committee which would hear petitions for and against the project before publishing its own recommendations. The tunnel's opponents again seized the chance to make their voices heard. By the end of the month, the work of the committee was in danger of seizing up completely as it tried to deal with a tidal wave of petitions, nearly 5,000 of them. Their number even included submissions from the Calais and Boulogne Chambers of Commerce. The magazine *PR Week* described it as the most intensive lobbying campaign for many years. Flexilink and the ferry companies were delighted.

At the same time, an opinion poll carried out for the European Commission suggested that public opinion in this country was swinging against the tunnel. It said that almost half of all Britons were opposed to the link and just three out of ten were in favour. The government was becoming concerned. It still seemed unthinkable that the tunnel's opponents could halt the project entirely, but they could damage it.

Much to the disgust of Aitken and his colleagues, Select Committee chairman Alex Fletcher MP decided to hear similar objections in groups, so that the committee would not have to hear the same arguments over and over again from different people. Even so, it looked as though the committee would have to sit well into the autumn before being able to produce its report.

Eurotunnel, though, remained sanguine. The Bill was intended to receive Royal Assent in April 1987. As long as the committee wound up its work in October 1986, then construction proper could begin in June as planned. "In terms of our timescale, we have got a bit of slack," said a laid-back spokesman.

With the tunnel again facing the prospect of a parliamentary delay in the UK, the project's French supporters must have looked on in amazement.

Outside the Channel ports, in France the project had overwhelming political support and the legislation was easily on track for completion in time for construction to start in mid-1987. There were two legislative processes going on at the same time. The French Parliament had to agree to the ratification of the treaty with Britain and to accept the concession agreement between the two governments and Eurotunnel. To that end, the relevant government departments drew up legislation, passed it to the Conseil des Ministres for approval, and then put it on the agenda for the National Assembly, the French Parliament.

The other process, called a Déclaration d'Utilité Publique, considered how government organisations and local people might be affected by the project. The DUP is a decree issued by the Prime Minister which gives the concessionaires the powers necessary to proceed with the scheme, such as the right to acquire land. Before the Prime Minister can issue the decree, he has to take advice from the Conseil d'Etat, which in turn must have studied three documents: a public enquiry report, similar to select committee reports in Britain, which focuses on private individuals and their interests; a study into how state organisations and government departments will be affected by the project; and details from the local councils affected on the changes needed to their land-use plans.

Throughout 1986, the French legislation progressed without a hitch. If there were to be any nasty surprises, there was no sign of them yet. Of course, there was overwhelming support for the Channel Tunnel in the British Parliament, too. But somehow such things are not so simple at Westminster.

While the politicians wrangled, Eurotunnel was working hard to raise the £6 billion it said it would need for construction, financing costs and a contingency reserve fund. The idea was to raise £1 billion in equity and £5 billion in bank loans. The equity would be raised in three stages. The first, called Equity One, would be worth £50 million. This would be "seed" money and would be put in by the founder shareholders of CTG and France-Manche, the ten construction firms and the five banks.

Equity Two would consist of a worldwide private placement of shares to raise £200 million. The timing of Equity One was flexible: the money was simply coming from people who were already committed to the tunnel and could be summoned at relatively short notice. But Equity Two would have to be in place by the summer of 1986 or even earlier if possible, to ensure that the contractor TML had enough money to pay its bills. Any delay might mean that surveying and design work would have to be slowed down.

The third and biggest chunk of equity, £750 million worth, was to be raised through a public share issue in July 1987. By this time, Eurotunnel would also have negotiated a £5 billion bank loan, although it would not be able to draw down any of that money until Equity Three had been completed. The loan could then either be used to pay for the construction work directly, Eurotunnel explained, or to support alternative methods of funding if these proved cheaper.

It was a complex and lengthy process. On many occasions, the fund-raising process was to prove as dramatic as the construction itself. Several times the project looked to be in serious trouble, yet each time it came back from the brink, apparently stronger than before.

That raising the funds should have been so nerve-wracking for those involved is hardly surprising with hindsight. The concept of privately financed infrastructure was in its infancy, not just in Britain but worldwide. Without government guarantees of any sort, investors were being asked to plough billions of pounds – or francs or yen or dollars – into a company with no commercial track record and into a piece of infrastructure that did not even exist, on the basis that cross-Channel traffic levels would boom at some time in the future. What is more, the construction industry had a reputation of going well over time and budget, especially on big projects – and this was the biggest of them all.

Convincing hard-nosed investors to back the tunnel was always going to be a struggle. By mid-1986, there were already signs that Eurotunnel's tight financial timetable was in as much trouble as the political one in Westminster. By July 1986, both Equity One and Equity Two should have been in place. Neither was. On August 14, Eurotunnel announced that Equity One had been signed. As of September 1 1986 it would have £41 million of new equity capital from its 15 founder shareholders plus a £9 million bank loan. But Eurotunnel had already been forced to push back the £200 million private placement – Equity Two – to the autumn. Rumours had been circulating in the City that there would be a delay and the news was confirmed on July 11 in London by Lord Pennock.

The Eurotunnel joint-chairman said he regarded the delay as unimportant and denied that it would affect the much larger Equity Three share offering planned for the summer of the following year. One of the problems was that the banks wanted more information on the construction contract, which at that stage had still not been signed, he said. John Franklin, a director of merchant bank Morgan Grenfell, which was advising Eurotunnel, was even more forthright: "There is absolutely nothing sinister about the delay at all," he said, and he saw no reason why the mid-1987 construction starting date should be affected.

He gave three key reasons for the delay. The most important was that Eurotunnel wanted to complete its negotiations with the banks over its crucial £5 billion loan before pressing ahead.

Secondly, it also wanted to reach agreement with British Rail and SNCF over what they would pay for using the tunnel, a key element in Eurotunnel's future income. And finally, Eurotunnel's financial experts were worried at the jittery state of the stock market and the number of new share issues planned for the near future. The conclusion was clear: far better to wait for improved conditions than to rush and risk defeat.

The decision may have been logical, but at the very least the delay did not look like a good omen.

Eurotunnel did deflect some of the criticism by announcing a series of high-powered additions to its board. They included Michael Julien, Midland Bank's group financial director; Sir Alistair Frame, chairman of the RTZ mining group and former head of the Channel Tunnel project which was abandoned in 1975; and most important of all, André Bénard, a former president of Shell France and a senior adviser to the Lazard Frères bank in New York. Bénard was to become one of the key people responsible for seeing the project through to completion.

But the delay only served to increase the pressure on Lord Pennock still further. He was an experienced industrialist, but some commentators began to ask whether the project would benefit from a change at the top. The *Sunday Times* was one of them. What about a dose of true entrepreneurial spirit in the form of Sir Nigel Broackes or Richard Branson, it asked? Without it, the paper continued, the project might remain what it has been since Napoleonic times, a dream on a sheet of paper.

Worries there may have been, but the pace did not slow. Eurotunnel had long claimed that the details of the construction contract had been agreed for some time. This latest hold-up had come about because it had to be approved by each of the 38 banks then backing the project. "It is only a paper delay," insisted a spokesman in July. "The construction firms know it is going to be signed and can get on with their preparatory work accordingly." And so it proved. It was signed on August 13 1986 in Paris, the day before Equity One was announced. The British half of the contract was worth £1.38 billion and the French half FF12.21 billion at 1985 levels.

The work was divided into three categories: target works, lump sum works and procurement items. The target works section covered the tunnels, which would be built on what is called a target cost basis. This meant that the contractor would be reimbursed for its actual costs, and in addition would receive a fee based upon the tunnels' target cost.

When the tunnelling was complete, if the actual costs were less than the target costs then the contractor would also take half of the money saved. But if the actual cost was more than the target cost then the contractor would have to pay 30 per cent of the extra out of its own pocket, up to a maximum of 6 per cent of target cost. The value of this part of the contract – in other words the target cost plus fee – was £699 million plus FF5,864 million.

The lump sum part of the contract covered the terminals and their related infrastructure plus mechanical and electrical work for the tunnels and all the fixed equipment, such as the track.

The agreed price for these works was £568 million plus FF5,024 million.

Finally, there were the procurement items. The idea was that TML would put certain items such as the locomotives and the shuttle trains out to tender, and that the bill plus an administration fee would be paid by Eurotunnel. The contract estimated this would cost £113 million plus FF1,322 million.

A lot more was to be heard about the contract in the years ahead. It was to provide the basis for just about every one of the many rows between Eurotunnel and TML. Many observers felt that when they signed this contract, the two protagonists sowed the seeds of all the problems that were to come.

The same week as the contract was signed, Eurotunnel appointed the firms of consulting engineers who would work as the job's Maître d'Oeuvre. The MdO acts as an independent project manager responsible to the client, in this case Eurotunnel, for the control of the project. It was a pivotal task for it would be up to the MdO to stop the inevitable rows from getting out of hand and jeopardising the project's future.

A joint venture of consultants W S Atkins from England and SETEC from France won the job, which would also involve reporting on progress or otherwise to the funding banks and to a special committee, called the Intergovernmental Commission, set up by both governments to oversee the operation of the concession agreement and, in particular, safety matters.

The MdO's appointment and the signing of the construction contract led to a burst of activity from TML. By November 1986, Glasgow-based engineering firm James Howden had won a £6.4 million order to supply the two tunnel boring machines needed for the British service tunnel – both would start from a cavern beneath Shakespeare Cliff, one heading out to sea and the other inland to the Cheriton terminal. The TBMs were to be delivered to the site in 70 loads weighing 20 tonnes each, and the first had to be installed and commissioned by December 1 1987 (which it was).

Leeds-based Hunslet Engineering was awarded a £1.2 million contract to supply construction locomotives and rolling stock, again for the UK tunnels. These would be used to take workers and materials into the tunnels and bring spoil out. In addition, TML had awarded site investigation and marine survey contracts worth roughly £3.5 million in total to Wimpey, Taylor Woodrow subsidiary Terresearch, Hydraulic Research and French firm Geocean.

If work in Kent was accelerating, on the French side it was approaching top speed. The 1975 works at Shakespeare Cliff had given the British contractors a head start. The French needed to sink a huge shaft at Sangatte outside Calais before they could even start tunnelling. The reason was simple. Most of the tunnel was to be built in a layer of waterproof chalk which runs roughly 50 metres under the sea bed. On the British side, this layer of chalk rises towards the surface at the coast, hence the white cliffs of Dover. Access to it was therefore fairly easy, but on the French side that layer of chalk does not rise. So, the contractors had to dig down to reach it. What is more, when the TBMs did finally head off towards Folkestone, they knew that they would hit very wet, fissured ground just off the French coast. So progress, at least at first, would be very slow.

As a result, the French were aiming at having their site fully operational by January 1987, well before the legislation would be passed in either country, and hoped to hit the layer of chalk by late spring. Tunnelling proper would start shortly afterwards.

The construction managers and site workers could hardly wait.

Speeding up the construction work was possible because Eurotunnel's finances were improving by the day. On September 25 1986, it received confirmation in writing from 40 international banks of their commitment to underwrite £5 billion worth of loans over 18 years. Bankers said this was this biggest and longest-lasting project financing deal ever arranged. Agreement was crucial if Eurotunnel's £206 million Equity Two private share placement, now set for October 1986, was to succeed.

Eurotunnel's standing was also boosted by the news that it had agreed the membership and structure of its board. André Bénard, who had joined the board in July, had become joint chairman with Lord Pennock. Others included Frank Gibb, the chairman and chief executive of Taylor Woodrow; Sir Alistair Frame, chairman of RTZ; Denis Child, deputy group chief executive of National Westminster Bank; Jean-Paul Parayre, chief executive of French contractor Dumez and the former president of France-Manche; Bernard Thiolon, chief executive of Crédit Lyonnais; and Sir Nicholas Henderson, former chairman of the Channel Tunnel Group. This was a powerful team by any standards.

At the same time, Eurotunnel appointed a chief and a deputy chief executive. The chief executive was Jean-Loup Dherse, a former main board member of RTZ. For the past three years he had been a vice president for the World Bank in Washington. His deputy was the former Midland Bank group finance director Michael Julien who, like André Bénard, had joined Eurotunnel two months before.

On October 29 1986, Eurotunnel announced that the delayed private share placement, Equity Two, had been completed successfully, raising £206 million. The shares had been placed not just in the UK and France, but also in the USA, Japan, Germany and Belgium. Lord Pennock, who had had a difficult year, was clearly relieved: "This is a very important step for Eurotunnel, one we always knew might be difficult before the ratification of the treaty." The project was back on track to open in 1993, he said. But although the placement was eventually successful, the run-up to it had been far from trouble-free.

The Channel Tunnel's long-standing opponents weighed into the debate. Flexilink, the ferry and port firms' lobby group, published its own financial assessment of Eurotunnel's prospects and argued that investors could easily end up with a nil return on their investment. Chairman Jonathan Sloggett explained: "There are three factors which are central to the success or failure of a Channel Tunnel scheme. These are construction overruns, traffic carryings and tariff levels."

The "nil return" scenario produced by Flexilink assumed construction finishing a year late and 12.5% over budget, with traffic levels growing by 1.95% each year but tariff levels falling by 40 per cent due to the extra capacity for cross-Channel traffic. Sloggett added: "Our consultants have demonstrated that the Eurotunnel financial case...is extremely sensitive to even minor variations in these important elements. On one set of calculations it has been demonstrated that Eurotunnel's total debt could rise to over £11 billion, their last loan would not be repaid until after 2023 and the return to investors would be zero. Eurotunnel's sums just do not add up."

Sealink's James Sherwood and MP Jonathan Aitken's group of anti-Channel Tunnel conservatives again all raised their voices in protest, agreeing that investing in Eurotunnel could be disastrous for shareholders.

For once, the doubts were not just being voiced by those with a vested interest in stopping the project. Many in the City of London were also nervous about investing, even though they had now seen details of the tunnel costings contained in the construction contract. One of the most outspoken voices was Richard Hannah, the respected transport analyst with stockbrokers UBS Phillips and Drew and a long-standing "chunnel watcher". "The key question is whether the likely returns will be sufficient to compensate for the risks involved," he wrote in a circular published in the run-up to Equity Two. Eurotunnel's base case model gives returns of around 17 per cent per annum over the lifetime of the scheme. Even if the base case assumptions are right, these returns are low by venture capital standards and, indeed, are below the UK industrial average in the last 10 years."

Given the political and financial obstacles that Eurotunnel still had to overcome before construction could start, investing in the company hardly seemed an enticing prospect. And when construction did start, there was always the danger that costs would rise sharply. Hannah pointed out that, in real terms, the costs of the NatWest Tower in the City of London almost doubled during construction and the cost of the Thames Barrier went up almost five times. "The construction environment has improved in the last 10 years, with lower inflation rates and better labour force agreements. Nevertheless, capital expenditure overruns are a major risk." Hannah concluded, rather wryly: "The City is often accused of being too short-term in its outlook and undoubtedly equity opportunities already exist with similar or better projected returns at a much lower risk. This unique offer will give the long-term investors the chance to stand up and be counted."

In an interview with *Construction News*, Hannah conceded that the tunnel was an unusual project: "It goes from being a very high risk venture, as it is at present, to being very low risk when construction is complete." But he stressed: "However, I do not see why it should be offering investors a lower return than can be obtained in other areas."

Hannah was not alone. An informal survey carried out by the *Financial Times* claimed that only six of the 25 largest pension funds and insurance companies were prepared to invest.

Eurotunnel's bankers hit back at Hannah's calculations. Dan White, a director of investment bank County NatWest, claimed that Eurotunnel's assumptions of a 17.4 per cent return on equity was very conservative, and he said that many income-earning possibilities, such as advertising, had not been included. He acknowledged Hannah's argument that the UK equity market had offered a higher rate of return – 22 per cent a year – over the last decade. "But if you take the last 55 years, which is the same period as the length of Eurotunnel's concession, then the rate of return is 6 per cent , quite some difference," White added.

The deadline for completing the Equity Two placement was 2 o'clock in the afternoon on Wednesday October 29. Britain and France were supposed to raise £70 million each, with £60 million from the rest of the world, mainly Japan, the USA and Germany.

The media was awash with stories that the British and American financial institutions were shying away from the project, leaving it to the Japanese and the French. The British government was rumoured to be lobbying hard in the City of London – and some reports even claimed that it had asked Washington to do the same on Wall Street.

The weekend before the deadline all the funds were rumoured to have been raised, apart from those of the British, who were £10 million short (something of a paradox as the project was privately financed at the British government's insistence). It was a tense time, but the money was raised. And with just Equity Three to go in nine months' time, the finance was almost in place.

7 Planning and designs

"It was a shock to us that they were not prepared to wait nine months when the funding banks were prepared to look forward 18-20 years."

Colin Stannard, National Westminster Bank

As it was struggling to raise the £206 million second tranche of equity in the autumn of 1986, the Channel Tunnel was also in the midst of its first major environmental row. If there had to be a Channel Tunnel, the green lobby was right behind the idea of an all-rail link, as it would produce far less road traffic than its drive-through rivals. The trouble was, how to dispose of the huge quantities of chalk, mud and other rubble that digging 150 km of tunnels would produce. At Sangatte, the problem was solved quite easily: a suitable land dump was found near to the construction site. But as usual, things were more complicated in Britain. At Shakespeare Cliff there was very little space and such a suitable site simply did not exist nearby.

In fact, about two-thirds of the British spoil could be disposed of easily – it would be needed for the construction work. The three tunnels would produce about 4.75 million cubic metres of spoil on the British side. (This compared with around 3 million cubic metres on the French side – because the ground was so bad off the French coast, the French TBMs would not get as far as their British counterparts before meeting up.) Around one million cubic metres was needed for fill at the Cheriton terminal site near Folkestone; another 1.9 million were required to build the working platforms on the Kent coast. It was the remaining 1.8 million cubic metres that the arguments were about.

Eurotunnel's idea was to use the spoil to provide a flat area of land at the foot of the cliffs which would be landscaped and used for "recreational pursuits such as fishing, walking and picnicking". Eurotunnel also argued that moving the spoil anywhere else would create a huge logistical problem and local people would have to put up with an awful lot of extra road or rail traffic.

Environmentalists reacted with horror. Dumping the spoil on the coast would cause unnecessary damage to the flora and fauna on the sea shore and also ruin a famous view from the cliff tops, said Robin Grove-White, director of the Council for the Protection of Rural England.

When the Select Committee of MPs studying the Channel Tunnel Bill restarted its work after the summer recess, the government announced that it had decided to back Eurotunnel, after looking at 70 alternative off-site locations, seven of them in detail.

The experts involved concluded that the problems of moving 1.8 million cubic metres of surplus spoil by road or rail to other sites would far outweigh any environmental gains at Shakespeare Cliff.

Another major concern was clearly that the search for a new site for the spoil could delay the project still further. Speaking for the Department of Transport, Michael Fitzgerald QC told the committee: "The government is anxious that there should not be a prolonged period of uncertainty...and has come to the view that no clearly preferred alternative had emerged." And for Eurotunnel, John Drinkwater QC said that even the cheapest of the alternative sites to Shakespeare Cliff would have added £40 million to the project's costs.

The Select Committee was not entirely convinced and asked the government to re-examine the use of alternative sites before the Bill's proceedings were completed. Chairman Alex Fletcher said: "We are concerned at the proposal for spoil disposal at Shakespeare Cliff."

But, if no better site was found, the committee simply decided to limit the amount of spoil that could be dumped there rather than ban dumping outright. Fletcher continued: "We require the Government to produce the text of an amendment...to limit the volume of spoil as cut to a maximum volume of 1.85 million cubic metres so that we may amend the Bill accordingly."

The green lobby kept a brave face. The CPRE said it still hoped to force the Bill to be amended further. But to everyone else, it looked as if Eurotunnel had won a decisive victory and another tricky and costly delay had been averted.

The Select Committee finally produced its report – having dealt with the 5,000 or so petitions – on November 18 1986. Almost all the major objections to the project and the Bill were overruled. Eurotunnel was delighted.

Admittedly, the report was rather late. Channel Tunnel supporters had originally hoped it would be produced before the summer recess, but the huge number of petitions put paid to that.

Eurotunnel was confident that the legislation could still be in place before the 1987 summer Parliamentary recess, but there was no slack at all left in the timetable now.

Apart from the spoil issue, the committee decided that the international passenger station should remain at Ashford; access to the Cheriton terminal should not be altered, as had been requested by Shepway District Council in Kent; the plans for the international passenger station at Waterloo and the North Pole railway depot in north London should remain in the Bill and therefore would not have to face a public inquiry; likewise, the planned improvements to the A20 should remain in the Bill and so would also avoid a public enquiry, although the MPs recommended that the line of the road might be changed to meet the needs of the petitioners concerned.

Not all the committee MPs agreed with the findings, however.

Labour MP Nick Raynsford produced a minority report in which he firmly aligned himself with anti-tunnel Tories such as Jonathan Aitken. Raynsford was critical of the hybrid bill procedure – as distinct from a public enquiry – which, he said, stifled opposition. "However hard the committee had tried to accommodate the petitioners, there is no doubt in my mind that many people have been left with the feeling that they have not had the opportunity to put their case fully and fairly," he said. "The committee has been subject to very considerable time pressures and has, in my view, consequently failed to take sufficient evidence on certain key issues."

Raynsford said he was particularly worried about safety in the case of a fire in the tunnel, and whether the risk of fires would be significantly reduced if passengers were to be separated from their vehicles. "In view of the horrendous potential consequences of a serious fire in the tunnel, it did not seem to be adequate simply to develop and test one approach, namely the preferred Eurotunnel option of keeping passengers in their vehicles during the crossing," he argued. "I believe that in the interests of safety...the alternative option of segregating passengers from vehicles should be evaluated. I cannot accept that safety be made subordinate to economic considerations and that the approach preferred by Eurotunnel should be the only one to be evaluated," he concluded.

Eurotunnel remained unperturbed at the one-man onslaught. "We do not consider it appropriate to comment on the decision of one member of the committee to publish a dissenting point of view," it said loftily. But it did reply to Raynsford's concerns about safety. Eurotunnel said it was confident that the independent experts appointed by the British and French governments to oversee the project would be satisfied with the safety arrangements. "We reaffirm our complete confidence in the safety of the system we are proposing."

Raynsford's minority report grabbed the headlines for a few days but had little effect beyond that.

The other MPs on the Select Committee had made 70 amendments but they mainly tinkered with the Bill. Nothing fundamental had been changed. The Bill could now move on to the Commons Standing

Committee and then the House of Lords, and the Channel Tunnel moved another small step towards fruition.

By the end of the year, Eurotunnel directors had revealed just how much difficulty it had had convincing the City of London to back its £206 million Equity Two private placement in October.

Colin Stannard, a senior banker then on secondment from National Westminster Bank, said Eurotunnel had been shocked at the reluctance of British institutional investors to back the project. He said it was astounding that the merchant banks, who advised the pension funds to stay away, would not consider an unquoted investment, even though a full quotation would follow in less than a year. “It was a shock to us that they were not prepared to wait nine months when the funding banks were prepared to look forward 18-20 years.”

He said that some British institutions, especially the insurance companies, did subscribe to the placement and when they did the overseas markets followed. The Americans put in £15 million and the Japanese £18 million, but there was still a shortfall which the founder members of Eurotunnel had to make up out of their own pockets. Stannard described it as “not that great”, and it was later estimated to be around £12 million. The experience, he said, illustrated the merchant banks’ belief in capital gain – in other words making a quick killing – rather than capital growth.

In the minds of many commentators, the problems Eurotunnel had in raising Equity Two raised serious questions about whether it could successfully raise Equity Three, which would be almost four times the size. But Stannard insisted that the worst was over. “I think it will be far easier to raise Equity Three than Equity Two,” he said at the time. Pension fund opposition would not be so strong because by the time Eurotunnel went public it would have a quoted share price.

During Equity Two, both American and Japanese institutional investors had expressed worries that a change of government at Westminster could scupper the project. Stannard said the Japanese were particularly wary of political uncertainty: “The Japanese have always appreciated the value of capital growth, but they won’t take a political punt.” The picture would be different by mid-1987, he argued: “The fact that a national law and an international treaty should be in place by the time of the stock market listing should quell any such fears permanently.”

Equity Three would be bigger. But Equity Two was the more difficult hurdle, argued the tunnel’s supporters. And that had been surmounted – just.

Meanwhile the contractors were pressing on. If they were to build the tunnel by 1993, they could not afford to wait for the legislation or the finance to be in place, even if this meant taking risks. At the end of October 1986 they started constructing the factory that was to build the concrete tunnel linings for the UK operation on the Isle of Grain in Kent. The linings would need huge supplies of aggregates and cement, and letters of intent for contracts worth £6.2 million and £9 million were sent to Foster Yeoman and Blue Circle respectively. To keep the road traffic to a minimum, all the materials had to be delivered to the Isle of Grain by sea and the linings moved to Shakespeare Cliff by rail.

TML also stepped up its recruitment drive, with a UK advertising campaign which attracted 10,000 applicants for just 500 jobs. And at the same time, the Department of Trade and Industry launched a campaign intended to help British suppliers, especially small and medium-sized ones, bid for and win work on the project (the initiative proved so successful that it is still running today, helping firms win work on major construction projects around the world, such as the new Hong Kong airport and the Storebaelt bridge and tunnel link in Denmark).

By the first week of March 1987, TML had placed, or was preparing to put out to tender, orders for tens of millions of pounds worth of services and equipment.

Consulting engineers Mott, Hay and Anderson won the principal design contract, worth £5 million. The firm had been the main designer on the previous scheme in 1975. Also in the pipeline were orders for around £25 million worth of steel reinforcement bar; 12 construction locomotives worth £1.5 million; mould and production line equipment worth £7 million; cast iron tunnel linings worth £9 million; roadheaders worth £2.4 million; £1.9 million worth of site accommodation...the list was huge. In all, £54 million worth of orders were placed or put out to tender in the first quarter of 1987, taking the total value of orders announced so far to £93.5 million, and the main construction work had not yet even started.

Eurotunnel calculated that for every million pounds worth of orders placed, roughly 30 jobs would be created or protected.

There was more to this, of course, than good planning by the contractors.

Towards the end of 1986, Prime Minister Margaret Thatcher had hinted that she might call an early general election, possibly in the spring of 1987. If the Tories won, Channel Tunnel supporters hoped that the Bill could be carried over into the first session of Parliament after the election. But if Labour won, the party had said it would expect the present legislation to lapse while it held a public enquiry or some other detailed investigation into the project. This could delay the project for months or even years, and could easily destroy Eurotunnel's complex financial plans. In such circumstances a future Labour government could well face the choice of backing the Channel Tunnel with public money or scrapping it altogether. The construction firms began to get an uneasy sense of déja vu.

Another Labour government; another project scrapped just as tunnelling was about to start.

Eurotunnel's insurance against these political uncertainties was to create as many jobs as possible before the election, so making it as difficult as possible for an incoming government to stop or delay the project. By now, TML had given orders to firms in Glasgow, Leeds, Norwich and Hartlepool, as well as Kent and the rest of the southeast. There would be many more in the months ahead.

When the various groups were bidding for the mandate in 1985, each boasted about the numbers of jobs they would create and that those jobs would be spread around the country. There was undoubtedly a lot of hyperbole in those claims, but now the scale of the project began to hit home. Whatever the precise figures would turn out to be, it would create many, many thousands of jobs. Indeed, with construction output in the UK growing steadily, the contractors were already becoming concerned at possible skill and labour shortages. To counter this on the British side, TML set up a training scheme with Kent County Council and the Manpower Services Commission to "prepare men and women for the variety of jobs that will occur as a result of the project". TML's concerns were not fanciful. With Sizewell B nuclear power station and Canary Wharf in London's Docklands under construction at the same time, the prospect of shortages was real.

The MSC put £100,000 into the scheme, which was also backed by the Construction Industry Training Board. The idea was simply to look at the skills and numbers of people the Channel Tunnel contractors would need and then, if necessary, to set up courses to make sure those skills were met. Early in 1987 it was too soon to say how severe the problems would be. "But it would be irresponsible for us not to recognise that there could be a lot of competition for skilled labour," said the TML director responsible, David Staines.

Finding the necessary skilled labour and at the same time keeping costs down as the construction industry entered its biggest boom since the last war was to become a recurring theme of the next three years or so, as TML scoured the world looking for experienced engineers and tunnel miners.

The year 1987 would be make-or-break for the Channel Tunnel. Either it would be scrapped for the second time in 12 years, or by December tunnelling would be under way on the most exciting civil engineering project this century. This was the year when the legislation was to be passed on both sides of the Channel; when the final tranche of finance was to be raised; and when the serious construction work was finally to start.

It began well. On January 22 the Bill finished its committee stage in the House of Commons and passed to the Lords. Eurotunnel had received something of a fright the week before, however. Labour MP Nick Raynsford had been an eloquent thorn in Eurotunnel's side, especially over safety, ever since he published his minority report following the deliberations of the Channel Tunnel Bill Select Committee. Raynsford was concerned that allowing drivers to travel with their vehicles through the tunnel presented an unacceptable fire risk, and that fires in tunnels, once started, can be very difficult to control. Drivers and their vehicles were segregated on the ferries, he pointed out.

Eurotunnel responded by saying that the shuttle trains would be provided with fire doors or curtains to prevent the spread of fire from one wagon to another, and that individual wagons could be isolated if necessary. Every shuttle wagon would have its own ventilation system to prevent the build-up of petroleum vapour.

And if a serious incident did occur, then the train could be stopped and the passengers evacuated into the central service tunnel where they would be rescued by special vehicles.

Raynsford was unconvinced. He proposed an amendment to the Bill which would have forced Eurotunnel to produce plans to segregate drivers and their vehicles. The Channel Tunnel Safety Authority – part of the Intergovernmental Commission set up by Britain and France to oversee the tunnel – would then choose the safest option.

If introduced, such a change could alter the whole rationale of the project. Costs would go up and Eurotunnel's service would look little different from that offered by the ferries or hovercrafts. Raynsford's amendment was defeated. But only by one vote.

Within weeks, safety was off the agenda entirely and the idea of crossing the Channel in a tunnel looked very safe indeed: the Zeebrugge ferry disaster, in which 188 people lost their lives, saw to that. To its credit, Eurotunnel refused to exploit the tragedy, but the argument that ferries were safe and tunnels were dangerous was heard no longer.

At the end of October 1986, Eurotunnel co-chairmen André Bénard and Lord Pennock invited Sir Nigel Broackes, chairman of Trafalgar House, to join the board. Lord Pennock's future with Eurotunnel had already been the subject of much speculation in the press, and Sir Nigel's arrival only added to it. The feeling was that he would take over as the British co-chairman of Eurotunnel sometime early in the New Year.

By February this strategy had collapsed. The contractor TML was wary of Sir Nigel's involvement in the project. He had led EuroRoute, a group led by Trafalgar House and British Steel, which had lost the race for the Channel Tunnel mandate a year before. Now they feared he was trying to win a slice of the work by the back door.

Just as damaging was what appeared to be his increasingly bitter relationship with Lord Pennock. In mid-February, Sir Nigel resigned and

Lord Pennock had let it be known that he too would be leaving when the time was right.

Banks and shareholders reacted remarkably calmly to the news. The timing was not good, with Equity Three on the horizon, but others, such as the highly respected Michael Julien, were still in place. The question was, who would take over? Speculation was rife. Perhaps Sir Nicholas Henderson would return. What about Cecil Parkinson, Sir John Harvey-Jones who was about to retire from the chair at ICI, or Cliff Chetwood, the head of the construction group Wimpey, which was one of the tunnel builders?

In the event the search did not last long. But the appointment was, at first, something of an anticlimax for Eurotunnel's new British figurehead was hardly a household name. Not yet, anyway.

Alastair Morton, who at 49 was younger than his rivals, was chief executive of the banking and financial group Guinness Peat, where he had been since 1982. Before that he was managing director of the British National Oil Corporation and was a government appointment to the board of British Steel between 1979 and 1982.

He was not a Harvey-Jones, the sort of high-profile figure that some were looking for to lead Eurotunnel through the trials and tribulations of Equity Three in the summer. *Construction News* carried his appointment in two paragraphs at the end of a story about skills shortages. Yet his arrival at Eurotunnel was one of the most significant events in the history of the project.

8 Morton takes charge

"Eat your heart out Heathrow."

Alastair Morton

Two days before Morton's appointment was announced, Eurotunnel suffered what in the City of London's eyes was its most serious blow yet – the resignation of its highly-regarded deputy chief executive Michael Julien, who was to join the troubled drinks group Guinness as finance director.

Julien was rumoured to have been unsettled for some time, and Eurotunnel was able quickly to make a number of high-profile non-executive appointments to the board. One was Sir Kit McMahon, the Midland Bank chief executive, former deputy governor of the Bank of England and a good friend of Morton. Another of similar calibre was Renaud de la Genière, a former governor of the Banque de France and now chairman and chief executive of the Compagnie Financière de Suez and of Banque Indosuez.

Even so, Julien's departure was a setback. After Pennock and Broackes, he was the third top Eurotunnel name to stand down in a week, and several fund managers wondered whether Eurotunnel would now be able to raise its final tranche of equity.

Morton, though, seemed if anything to thrive on the difficulties. He quickly instilled vitality and confidence into the project. Previously even Michael Julien had been cautious about Eurotunnel's future prospects. The day after his departure he was quoted in the *Financial Times* as rating Eurotunnel's chances of completion at "well over 50%", but added that he could not be more precise "because there are other rocks in the way". Hardly an overwhelming vote of confidence. Now here was Morton full of enthusiasm, excited at the prospect of linking Britain to the rest of Europe and telling potential investors to put their hands in their pockets and back him.

Known as a combative individual, he tended to arouse fierce loyalty and dislike in equal measure. His negotiating skills were not in doubt. He had won BNOC a major North Sea oil stake from the last Labour government.

And although he had the reputation of disliking Margaret Thatcher, he was close to Nicholas Ridley's successor as Transport Secretary, John Moore, and had the backing of the Bank of England. Indeed, many believed the government, through the Bank of England, was instrumental in Morton's arrival.

The Observer newspaper described him as "an acknowledged Bank of England troubleshooter" and said the government had been looking for a replacement for Lord Pennock that it could trust to keep the project disaster-free in what was increasingly looking like a general election year.

Margaret Thatcher had pinned her faith on the abilities of the private sector. The Channel Tunnel had become a symbol of that belief. If it went into terminal decline, the political fall-out could be disastrous for the Tories. *The Observer* added: "Morton's very appointment is in some ways an embarrassment for the government. It is a confirmation that the Conservatives have been forced to assume direct control of their most ambitious private sector project."

But any concerns that Morton might be a government stooge were rapidly dispelled. From the first day he spoke his mind and was very much his own man. And he was tough with it, as the contractors and his own Eurotunnel colleagues were soon to discover.

An early example of his sharp-talking style came at a meeting of the Global Pension Fund Forum, a select group of around 100 institutional money managers whose support could be crucial for Equity Three. Talking about the problems of late 1986 and Equity Two, his new colleagues must have winced when he said: "The ten contractors and five banks who won the concession must take stick. It took them until September and October 1986 to start bringing an independent Eurotunnel management team together. Frankly, Eurotunnel only began to develop a mind and a will of its own as 1986 ended: no wonder it looked doubtful last October."

To many people it looked pretty doubtful in 1987 too, Morton at the helm or not.

Within weeks of starting work, the new co-chairman and his colleagues had decided that the £750 million share issue planned for July would have to be delayed until the autumn. Equity Three was crucial. If it failed there would be no Channel Tunnel for a couple of decades at least. But delays in the British Parliament, the management upheavals at Eurotunnel and the prospect of a June general election – now the hot favourite, although there had been no official announcement – all meant that sticking to the original timetable would have been a very high-risk strategy. Eurotunnel had enough trouble convincing British investors to back the project already without making things unnecessarily difficult. Morton was later to say that this was a crucial decision.

If construction work was to keep going, however, Eurotunnel would need at least enough money to pay TML, so it decided to raise £75 million with what it called a "mini-equity" launch in the summer to tide it over. TML was told to keep to its original construction programme and the move was accepted as sensible in financial circles. It also gave Morton and Bénard time to concentrate on two other key issues – the next stage of the £5 billion bank loan agreement and the contract with the railways.

Reaching a deal with British Rail and SNCF over how much they would use the tunnel and what they would pay for the privilege should have been straightforward. After all, they were both big supporters of a rail tunnel which would give them a competitive edge over their road-haulage rivals.

A deal should have been signed by the end of 1986, but with Easter 1987 approaching the negotiations were dragging on. Morton, in his inimitable style, decided to shake things up a bit. He let it be known that the stalled negotiations with the railways could lead to the entire project's collapse. Eurotunnel sources said they were alarmed at the railways' apparent lack of commitment, and in particular at British Rail's decision to run down some of its inland freight depots.

The core of the dispute was that the railways were taking half of the tunnel's capacity but only paying around 40 per cent of the revenue. Eurotunnel wanted more money and it wanted most of it paid in advance.

There was also a dispute about the speed of the trains. The shuttles, owned by Eurotunnel, would carry vehicles through the tunnel at 100 mph. The railways wanted the through trains to go at 125 mph. One would have to slow down or the other speed up.

British Rail did not rise to Morton's bait. A spokesman simply said: "We are only too aware of the importance of the negotiations and there is no lack of urgency on our part."

But three weeks later the deal had been agreed. It was finally signed on July 29 1987.

The contract gave the railways half the capacity of the tunnel. In return they agreed to pay usage charges to Eurotunnel, composed of a fixed annual amount and a variable amount which would depend on the number of passengers and the amount of freight using the system. They also agreed that the minimum they would pay Eurotunnel each month would be £9 million, at November 1987 prices, an arrangement that would last for 12 years.

Both sides were pleased. Eurotunnel had failed to raise the levels of tolls to be paid by the railways but the deal gave it an assured level of income. The railways had won the right to 50% of the tunnel's capacity, as they had originally wanted.

In a statement issued on the morning of May 12 1987, messrs Bénard and Morton said: "Eurotunnel has gained a much more certain basis for the project...our negotiations with the chairmen of British Railways and

SNCF have been friendly but robust." The agreement also confirmed one of the great advantages of the tunnel in the eyes of its supporters – the future ability of the railways to compete with the airlines on the short-haul European flights.

Morton was beginning to sound ebullient even by his standards. A few days later he told a meeting of financiers: "Suppose you are doing business here in London. You could leave your client lunch in Aldwych by 2.15, walk across the bridge to Waterloo, board the 2.30 to Paris and be stepping into a limo at the Gare du Nord before 6.30 local time (they are one hour ahead), fresh and ready for a 6.45 appointment at Crédit Lyonnais or BNP in the Boulevard Haussman, followed by a 7.45 clients' dinner nearby which you must leave at 9.30 to catch the 9.45 to London. Waterloo by 11.45 and home in Kensington or Chelsea by midnight, after a snooze or a useful session with your notes on the way back. Eat your heart out Heathrow."

The months of May and June 1987 brought further good news. The European Investment Bank said it would be lending Eurotunnel £1 billion, and the Select Committee in the House of Lords studying the Channel Tunnel Bill recommended no major changes. The Bill would now go before a Standing Committee, as it did when passing through the Commons.

With the general election now set for June 11 1987, the Bill looked certain to be ready for Royal Assent by the end of July – after the general election but, crucially, before the long summer recess. The only serious political question left was, what would happen if Neil Kinnock beat Margaret Thatcher?

It now seemed that only the dismal failure of the Equity Three share issue or a new Labour government sticking to its pledge to call a public enquiry could stop tunnelling proper starting before the end of the year. Labour would need an overall majority, too. The Alliance had announced that if it held the balance of power it would not try to halt the project and would not support calls for a public enquiry.

On June 4, the French Senate unanimously approved the legislation which would allow the government to ratify the treaty and the concession. The National Assembly had already given its unanimous approval on April 22, and the Déclaration d'Utilité Publique giving the go-ahead for construction itself had been issued in May. In France the political process was all but over.

On June 11, any remaining political uncertainties in the UK were removed when Margaret Thatcher was re-elected with a huge majority. Royal Assent for the Channel Tunnel Bill was on target for July.

Six days later in Paris, another problem was solved. Bénard and Morton announced that they had arranged an interim financing package of £72.5 million to tide Eurotunnel over until Equity Three in the autumn. However, instead of a mini share issue, ten banks close to the project offered a loan repayable out of the public issue. "Step by step, this great project is going into place," they said. It was hard to disagree.

The loan allowed the contractor Transmanche Link to place a new set of orders for construction materials and equipment, worth around £100 million on the British side alone. By the end of August 1987, TML had placed orders worth £30 million in total with Howden of Glasgow and Robbins–Markham (the joint venture of Robbins from America and Markham from Chesterfield) for the land and marine running tunnel boring machines; and a £2.3 million consultancy contract with the Building Design Partnership for terminal design.

It had also ordered £353,000 worth of hardcore, £359,000 worth of crushed concrete, a sewage plant, £2.1 million of pulverised fuel ash, plus lasers, transformers, rebar, payroll systems, forklift trucks, tractors, conveyors and hoppers, hammers...you name it, TML bought it.

By the end of July, the Sangatte shaft was complete at a cost of £17 million. Small roadheaders (a kind of tunnelling machine) were being used to start the tunnel off towards England, with the TBMs taking over once they had completed 60 metres.

At Shakespeare Cliff preparations were well under way for the start of tunnelling. The site had been prepared for the arrival of the first TBMs, with new access shafts and adits built to complement those constructed for the 1975 project and which were being reused.

By August 3, work was scheduled to finish on the first of the TBMs being built at Howden's Glasgow factory. The 4.8 metre diameter machine would then be dismantled, carried to Kent and reassembled in a specially constructed underground chamber. In November, it would then begin work on the stretch of the central service tunnel running 8 km inland from Shakespeare Cliff towards the terminal site. The TBM which would dig the service tunnel from the foot of Shakespeare Cliff out to sea would be delivered to site in November, ready to start work before the end of the year.

By this time, TML had also completed the last of the surveys of the tunnel's cross-Channel route and was satisfied there would be no unexpected surprises. The surveying process had been extensive. There were 100 boreholes already in existence. These were examined and 12 more were drilled 100 metres into the sea bed at a cost of around £500,000 each.

Like Eurotunnel, TML had in addition been strengthening its own management team, most notably by bringing in one of the senior figures of British construction, Andrew McDowall, to be the UK chairman of the TML supervisory board. The logic of his appointment was simple, said Sir Frank Gibb, chairman of TML consortium member Taylor Woodrow and one of the key figures on the project. He sounded as optimistic and enthusiastic as Morton talking to a meeting of sceptical bankers. "Good progress is being made in the development studies, the initial subcontracts have been placed and the preliminary site works have now begun...in anticipation of the main construction work starting later on this year. This build-up of construction activity has called for an increase in the top management team."

McDowall had an impressive track record as a director of Wimpey and managing director of George Wimpey International and Engineering. He had been responsible for some of the firm's largest overseas projects, most notably the prestigious new headquarters of the Hong Kong and Shanghai Bank. But like many of his contractor colleagues, by early 1988 McDowall must have wondered why he had accepted the job, as Morton and Bénard turned their attention away from the banks and pension funds and focused on what they saw as muddle and inefficiency within TML.

For the moment, though, everything seemed just fine. The project had made huge progress in 1987. The contractors were almost ready to start tunnelling. And Bénard, Morton and Eurotunnel were concentrating on how to win the battle of Equity Three.

The general election over, the Channel Tunnel Bill returned to Parliament and the formality of Royal Assent. But, as if to keep everyone on their toes, there was a last-minute hitch. The problem was amendments introduced in the House of Lords which changed the route of some of the tunnel's access roads. These amendments were technically not allowable, as the landowners affected by the new road routes should have been told 18 months beforehand. And the Commons Standing Orders Committee again caused Channel Tunnel investors and supporters palpitations by refusing to allow the amendments through.

The result was that the government had to introduce a motion to override the Standing Orders Committee's decision. Otherwise the Bill would not have reached the statute book before Parliament's summer recess, and the future of the whole project could again have been thrown into doubt. At the very least the opening date, and hence the date when Eurotunnel would start earning money to pay off its huge debts, would be put back.

That was the last Parliamentary problem, however. The Bill did go through and it received Royal Assent on July 26. Just four days later, the treaty was ratified by Margaret Thatcher and François Mitterrand. The political process was complete at long last.

At the end of August 1987 a new threat to the project briefly emerged, one that had hardly figured in any discussion about the tunnel to date: strikes among construction workers.

If this project had been being built during the 1960s and 1970s, the effect of industrial disputes would have been one of the hottest topics and one of the private sector's key worries when deciding whether to invest or not. Margaret Thatcher's arrival in Downing Street had changed the political landscape to the extent that analysts studying the project would point to the construction industry's improving industrial relations and productivity records.

Stockbroker Savory Milln's comments, contained in a circular on Eurotunnel published in the autumn of 1987, were not unusual: "The construction industry is much more efficient than it was in the seventies. For example, the real cost of constructing motorways has fallen, with some projects being completed a year before schedule. More demanding customers insisting on incentivised contracts have galvanised the industry by offering big financial rewards for completion within time and cost. Some of these incentives are passed on to the labour force which, with tougher management, has created much better industrial relations and made the UK industry unrecognisable from the bad old days."

Were those bad old days about to return, however? And how would the French unions react to the project?

Economists were predicting that the UK construction industry would grow by 4.5 per cent in 1987, the highest level of activity recorded since 1979. And that was leading to upward pressure on wages.

On the Isle of Grain in Kent, workers building the factories which would manufacture the tunnel linings walked out for a week.The strike started when 56 labourers, all members of the Transport and General Workers Union, downed tools in a dispute over bonus payments. They had been demanding a guaranteed minimum bonus payment of £2.50 an hour, to be paid on top of the basic rate of £2.26 an hour. The strike lasted six days and ended when TML management agreed to talks on the claim.

Two weeks before, French construction workers walked off site at Sangatte. That was settled even quicker, within two days, when TML offered the 151 workers a package including increased bonus payments and permanent contracts of employment.

According to the workers' union, the Confédération Général du Travail, the new bonus arrangements should add between 10 and 15 per cent to pay packets. But it added that more disruption could follow if talks on improving health and safety procedures were not concluded satisfactorily. And CGT secretary Robert Brun said that the issue of low pay had not gone away, as the increase would leave many workers on a basic wage near the French legal minimum.

Neither dispute was important in itself, but they served as a warning to potential investors and the TML management that good industrial relations were not to be taken for granted.

Eight months later, the threat of industrial unrest on the Channel Tunnel resurfaced when the TGWU threatened a strike of its members on major construction sites around Britain in support of a pay claim. That threat did not last long either. It was made in mid-May 1988, and a pay deal was signed by mid-June. In fact, such disputes, and even threats of them, were rare. Relations between unions and management were to remain good throughout. The problems of TML and Eurotunnel would lie with each other, not the construction workers.

By the last week in August 1987, Eurotunnel was trying to put in place the final pieces of its huge financing deal. The £5 billion loan from 50 banks worldwide was formally underwritten on August 24, with 12 Japanese banks providing £1.5 billion and five British and five French banks a further £1.3 billion. Later that week, Morton and Bénard left on a whirlwind worldwide tour to help the 50 banks so far involved syndicate the loan to a further 200 or so, thus spreading the bankers' risk. Typical of the pace of the project, it was a punishing schedule. Morton visited Toronto, Tokyo and New York, while Bénard went to Oslo, Milan, Madrid, Frankfurt, Zurich, Amsterdam and Bahrain.

The sight of the Eurotunnel tunnel co-chairmen circling the globe to attend one banking conference after another was eye-catching and headline-grabbing, but the loan negotiations were now under control.

What was less certain was the final piece of the financial jigsaw, the delayed Equity Three share issue, described by Richard Tomkins in the *Financial Times* as "an equity offering the like of which the public has never seen before".

He went on: "One of the many peculiarities of Eurotunnel is that it is among the biggest infrastructure projects of the century, yet it is being built without government money or even government guarantees on either side of the Channel. That is unusual for a project of such economic and political significance. It also means, of course, that it has to be sold to investors on the basis that it will deliver commercial returns – not an easy matter when it has yet even to be built, and when those returns can be based only on forecasts stretching far into the next century."

Eurotunnel intended Equity Three to raise around £750 million, but its significance was far bigger than the cash it would raise, for if it flopped, the banks had said their loan agreement would automatically fall. In other words, the Channel Tunnel would be stopped in its tracks.

In September 1987, the campaigning started in earnest, with the publication of the latest two out of a series of papers on the tunnel from Eurotunnel's lead broker, Warburg Securities. These two reports focused on the construction process itself and what Warburg called the investment case – in essence a summary of the reasons for backing the project. The latter included Eurotunnel's latest forecasts for cross-Channel traffic. In

1985 48.1 million passengers crossed the Channel: Eurotunnel was predicting that that figure would have increased to 111.9 million by 2013. Freight would increase at roughly the same rate.

These were huge rises, but other analysts supported the figures and even suggested that Eurotunnel might have been over-cautious.

In 1982, Sir Alec Cairncross wrote a report for the government on cross-Channel traffic which, he said, "has a way of mushrooming unexpectedly". Stockbroker Savory Milln agreed: "We are confident that Eurotunnel has not opened itself up to the accusation of being over-optimistic; indeed the contrary could be argued."

It did add a cautionary rider, however. "Eurotunnel's projections are sound but they should be regarded...as suggestions more than forecasts. They...are the best figures available but they are not designed to anticipate the unexpected."

Warburg's construction report was just as upbeat. "We believe the balance of probability is that Eurotunnel will be completed both on time – May 1993 – and to budget. Throughout our investigation of the Channel Tunnel project we have been impressed by the conservative civil engineering assumptions that Eurotunnel have included when calculating both the tunnel's cost and its completion date."

The report concedes that there were risks, especially in a project of this size, which "is open to very many external influences". But it added: "However, the extent of the research to which the proposed crossing has been subject, and the strength of the contracting team involved, provide great confidence. Most importantly we believe that the possibility of it being delivered either so late or so far over cost that it would seriously damage the viability of the project can almost certainly be discounted."

Despite the recent conflict at the Isle of Grain and Sangatte sites, the construction report's author, Phillip Raper, was confident that the project would not be dogged by poor industrial relations. He said that whatever disputes did occur, they would not affect the tunnel's viability as the unions on both sides of the Channel were behind it. He added that he reckoned there was a six-month slack in the timetable in case a serious dispute should develop. And he, like others before him, stressed that industrial relations in the construction industry had improved out of all recognition since the 1970s.

The big problems facing the contractors would be logistics – getting workers and materials into the tunnel and spoil out – and dealing with those parts of the project that were truly ground-breaking, such as the design of the shuttle wagons.

With everyone being so positive, it was hard to see what could possibly go wrong. But this enthusiasm was making many tunnelling experts uneasy. True, the Channel Tunnel was using tried and tested technology. True, it would be going through an ideal tunnel medium, at least for most of its route. And it would be going through one of the most heavily surveyed stretches of ground ever. But tunnelling projects always contained a large element of uncertainty and risk: you never really knew what the ground would be like until mining started.

In January 1986, just days after Eurotunnel – then called the Channel Tunnel Group and France-Manche – had won the mandate, *Construction News* commented that building the link was likely to be "far more complicated in practice than it is in theory". It added: "Even relatively short tunnels can produce major problems." One recent example was a water treatment tunnel at Iver in Buckinghamshire, which was going through what had been described as "extremely well-documented and well surveyed ground". The contractor Thyssen hit an unexpected underground well. It held up work for more than a year and £2 million was spent dealing with it before tunnelling could resume.

Another example, more on the scale of the Channel Tunnel, was the 34-mile-long Seikan Tunnel in Japan, the world's longest undersea tunnel. Breakthrough was in March 1985 and it opened to rail traffic three years later. It took over 20 years to complete and cost £3.7 billion, twice as long and ten times the cost in the original estimate. It also killed 34 construction workers.

The geology in Japan was much more difficult for tunnellers than that in the Channel – so much so that despite the problems the project could rightly be acclaimed as a triumph of civil engineering.

Four serious floods, one of which almost caused the entire tunnel to be abandoned, were overcome. The biggest test for the contractors came in 1976, when they ran into a massive fault zone which the pilot tunnel had failed to pick up. Some 1.4 km of the main tunnel was flooded as water flowed in at a huge rate, estimated at 43,000 litres a minute. The water was eventually contained, the fault sealed and that area of rock by-passed. But it served as a reminder that no matter how good the contractors, tunnelling is a risky business.

9 High risk, high return

". . . built upon an unstable pyramid of uncertainties."
Jonathan Sloggett, Chairman, Flexilink

The Channel Tunnel had so far had a good 1987, but its opponents were not defeated yet. With Equity Three looming, the ferries and Flexilink relaunched their anti-Eurotunnel campaign in an attempt to dissuade investors from backing the share sale, now set for November.

Sealink Chairman James Sherwood renewed his warning that 20,000 ferry workers could lose their jobs because of competition from the Channel Tunnel.

Flexilink chairman Jonathan Sloggett described the project as an imprudent, high-risk investment even for those investors seeking modest returns. He predicted that Eurotunnel would have to default on its loans and that this would lead to a "lengthy or even indefinite postponement of returns to equity holders". He added: "Eurotunnel itself admits that investment in the project 'involves a significant degree of risk', a view which Flexilink wholeheartedly endorses. Of great concern to potential investors must be the long wait for any dividend, the likelihood of adverse variations in costs and revenues and the long-term pressure for a competing drive-though scheme."

Eurotunnel's assumptions on the key variables – construction costs, traffic levels and pricing in a competitive environment - he argued, were either "fundamentally flawed or built upon an unstable pyramid of uncertainties".

Eurotunnel remained cool, knowing that this was Flexilink's last shot. Morton and Bénard's world tour – dubbed Eurotunnel's international roadshow – to help syndicate the banks' loans was winning friends for the project. That was more important than getting involved in a shouting match with the ferries.

Equity Three would be tough, but Flexilink had failed to harm the tunnel so far and there was no reason to suppose they could succeed now.

While Eurotunnel's top brass were travelling the world lobbying potential investors, the contractors were making what looked like very good progress, adding to the increasingly upbeat atmosphere that had developed since Morton's arrival.

The consortium of American firm Robbins and Chesterfield-based Markham had been awarded a £15 million contract to build two of the full-size tunnel boring machines needed for the British side.

These would be used to drive the main train tunnels towards France. The final two for the tunnels from Shakespeare Cliff back towards the terminal at Cheriton looked certain to be built by Howden of Glasgow. Howden had already won the job of building the two service tunnel TBMs and the first of these, 160 metres long and weighing 600 tonnes, had been completed and was being assembled underground beneath Shakespeare Cliff.

The service tunnel would begin at the site of the abandoned 1975 workings. The plug of the old tunnel had been removed and the British side of TML was optimistic that it could start tunnelling on December 1 1987. If it did, it would mean that the British had overtaken their French counterparts, whose TBMs were not expected to start work at Sangatte until December 6.

The British firms were clearly delighted at this small triumph over their French colleagues. "The French have received a lot of publicity, but the British have been quietly beavering away to good result," said one.

At the end of September, relations between Eurotunnel and TML abruptly changed for the worse. Eurotunnel chief executive Jean-Loup Dherse resigned to join the Vatican. His job was taken over by Eurotunnel managing director Pierre Durand-Rival, who came to the post with a tough reputation and wide experience of major construction projects, including the Solmer steelworks near Marseille. This reputation was immediately enhanced when he launched a stinging attack on the contractors' record to date – a record the contractors themselves were rather proud of. Durand-Rival wrote to TML criticising its "failure to comply with its obligations under the contract", and he told TML to sort out problems including "continuous delays" in the engineering programme and a "total lack of financial information".

He wrote: "Orders are placed and approvals for design are requested without any reference to budget. While such difficulties are not unusual at an early stage of a major project for an emerging contractors' organisation, we note with increasing concern your inability to overcome these difficulties."

Referring to the approaching Equity Three launch, he added: "At a time when Eurotunnel is about to make a major financial commitment, the present situation is clearly unacceptable and most disturbing."

Ironically, the letter reassured rather than worried City of London financial experts, who felt that the relationship between Eurotunnel and TML could become far too cosy to the detriment of shareholders.

Richard Hannah, experienced Chunnel-watcher and transport analyst with stockbroker UBS Phillips and Drew, said the letter raised considerable doubts about the construction programme and whether it was under control. "But it does demonstrate that Eurotunnel is becoming more independent and that is to be welcomed as far as the institutional investors are concerned."

Officially, TML did not comment on the letter which had been written to Andrew McDowall, chairman of the supervisory board.

Privately, reaction was divided. Some conceded that there were teething problems, but argued that that was inevitable on an international project of this scale. They felt that Eurotunnel was just showing potential investors that it was in control of the project rather than the construction firms. In that light, there was no reason to overreact.

Others, particularly those lower down the management ladder, were not so happy. The project had made huge strides in recent months. This was an unwarranted slur on their efforts and did nothing to keep morale high.

Whatever the reasons for writing it, Durand-Rival's letter was to prove the first shot in a very long war of words between contractor and client.

On November 4 1987, the final details of the £5 billion credit agreement, now syndicated to 198 banks worldwide, was signed simultaneously at London's Guildhall and at the Pavillon Gabriel in Paris.

Twelve days later, on November 16, Eurotunnel published its prospectus for Equity Three. At the same time it announced that the offer – 220 million units at £3.50 each to raise £770 million – had been fully underwritten. This meant that even if all the shares were not sold, the underwriters would be obliged to buy the balance.

The share offer would close at 10 in the morning on November 27, with 101 million on sale in the UK, the same number on sale in France, and the balance of 18 million being sold to international clients. Of the UK portion, 42 million had already been allocated to certain institutional investors and their clients. The balance of 59 million was on offer to the general public.

The British contractors had intended to start tunnelling from Shakespeare Cliff towards France on December 1. They would now be able to start as planned.

The prospectus itself contained few surprises. Eurotunnel had raised its cross-Channel traffic forecasts slightly, which would make the tunnel more appealing to investors. Its prediction of passenger trips in the year 2003 now stood at 94 million, compared with a forecast of 88 million for the

The magnitude of the undersea excavation stunned visitors to the tunnel site.

At the back of the cutting head of a tunnelling machine one of the 700,000 lining segments is fitted into place.

(Top) The segments which form the tunnel lining, cast in the strongest concrete in the world, wait their turn at the Isle of Grain.
(Bottom) The underground locomotive system left little room for the tunnellers.

(Top left and right) The tunnel miners stuck to their task to breakthrough on time despite early serious delays.
(Bottom left) Tunnellers were able to enjoy a break for tea even as they battled to make up time.
(Bottom right) Engineer Mark Dowell operates the laser system that ensured the tunnelling machines met under the sea.

same year given in July. Actual passenger trips in 1985 stood at 48 million.

It also forecast that Eurotunnel would take 44 per cent of this enlarged passenger market and 17 per cent of all freight traffic, equivalent to some 30 million passengers and 15 million tonnes of freight a year.

But the prospectus did reveal why Eurotunnel managing director Pierre Durand-Rival had been so upset two months earlier. The contractors had undoubtedly achieved a great deal, but nonetheless the project was already significantly behind schedule. "The UK tunnel linings plant is now in operation. A similar facility in France and the access shaft at Sangatte are nearing completion....Orders and subcontracts have been placed with a value of over £200 million and more than 1,800 people are currently engaged on the project." But it added: "Certain operations are behind schedule. The view of the Maître d'Oeuvre is that the project is currently about three months late and that the working programme would presently indicate a maximum delay of not more than five months in the completion of the system."

The MdO had apparently told Eurotunnel that the time lost could be recovered, and talks with the contractor about how best to achieve this were already under way.

Despite the construction delays and the stock market crash of "Black Monday" just a month before on October 19, the omens for Equity Three seemed fair. Alastair Morton was certainly optimistic. "This is an historic occasion. The equity capital and the loan finance are committed. The opportunity is now open to the public to join in this great venture and already over 550,000 have registered with our Share Information Office. With these levels of interest we can be confident that many members of the public are going to invest in Eurotunnel and I look forward to welcoming them, as our shareholders."

Press reaction was almost as positive, with some reports talking of a likely total of half a million share applicants in the UK and the possibility that the shares would enjoy a premium as high as 50 pence when dealings began on December 10.

It was not to be. As with Equity Two, the response was muted. In the UK, the much-touted total of 500,000 applicants never materialised: only 112,000 investors applied. They took 38 per cent of the stock on offer. Various institutions had already agreed to take a further 42 per cent, and that left the underwriters with the remaining 20 per cent.

Things were no better in France, where the issue was marred by a new wave of strike action at the Sangatte site by the pro-communist CGT union which was demanding better pay and other benefits.

Banque Indosuez, one of Eurotunnel's French advisers, said that the French public bought about 60 per cent of the stock and a further 15 per cent was purchased by institutional investors unconnected with the project. The remaining 25 per cent had been either bought by existing institutional shareholders or left with underwriters.

Typically, if Morton and Bénard were disappointed, they did not show it. "I don't regard the response as a setback or failure, I am extremely pleased," Morton said. Under the circumstances – a hugely ambitious project, a nervous market – he had the right to be.

Richard Tomkins of the *Financial Times* agreed. "If there has to be a final judgement of the flotation it will probably be seen as a minor triumph...the tunnel itself will in all probability now be built."

The story had one further twist. After their first day of dealings on the London and Paris stock markets, Eurotunnel shares closed 100p down on the offer price at £2.50. Not that this bothered the contractors much. The money was now in place and they had just under five and a half years to build and fit out 150 km of tunnels and two huge terminals.

10 The contractors start work

"What we have done here is revolutionise tunnelling."
Colin Kirkland, Chief Engineer, Eurotunnel

By early 1988, with construction gathering pace, it was now easy to believe that the tunnel would be built. But, in Britain, this raised a sensitive question. A Channel Tunnel was all well and good, but how would travellers get to it?

The French had already announced their intentions – a new coastal motorway, an extension of the Paris motorway into Calais and, of course, the high-speed train line linking Paris, Lille, the tunnel and Brussels.

So far, Britain had restricted itself to an extension of the M20 motorway into Folkestone, some relatively minor road improvements in Kent and promises from British Rail to spend £400 million improving the Kent railway services and building an international terminal at Waterloo.

Kent had always had something of a love-hate relationship with the Channel Tunnel. Inevitably, a great many of its inhabitants were strongly opposed to it and the upheaval it would bring.

Others feared that the tunnel would lead to mass unemployment in the ferry ports, with thousands of people who directly or indirectly relied on the ferries losing their livelihoods.

But Kent is not just a county full of ferries, retired colonels and orchards. The north and east of the county have unemployment almost as high as anywhere else in Britain. As far as this part of the county was concerned, if the Channel Tunnel brought jobs it would be welcomed with open arms.

Kent County Council was also possessed of a strong pragmatic streak. It believed early on that both British and French governments had the political will to see the tunnel built. And if it was going to be built then the people of Kent had the right to expect that the necessary transport links would be provided to ensure that travellers could cross the county as easily as possible. The key to a successful Channel Tunnel transport policy was an efficient rail link which would take as many cars and trucks off the roads as possible.

Needless to say, the need to develop efficient transport corridors for road and rail traffic had an enthusiastic supporter in Alastair Morton. The unmistakable signs of official and ministerial lethargy on the issue were worrying him. After all, as far as he was concerned, this was Eurotunnel's future income at stake.

He spelt out his fears in blunt terms at a railways conference in London in early May. He argued that the road system was sufficient, "subject to continuing attention to bottlenecks and weak links", but only as long as the rail and air networks played "very strong complementary roles". But air-space and runways are severely limited, so that means the train taking the strain. "But for former air travellers to accept that, speed, comfort and reliable frequency are essential. Similarly, speed and reliable frequency are essentials for former freight users to accept long-haul rail freight as being usable in practice as well as economical." He continued: "And there is another party for whom, I believe, it is vital that rail plays to the full its potential role in taking the growth that air and road cannot take. That party is Kent. Unless rail access to the tunnel is optimised around existing rail routes, Kent, to misquote Noel Coward, will be asked again and again 'to lie back and think of England' as yet another motorway is ripped across the county."

Morton was not alone in thinking Britain was lagging behind. The Confederation of British Industry, the trade unions, the National Economic Development Office, the banks and the construction industry all pressed the government to plan better links to the Channel Tunnel.

In France the plans were well in hand for the new high-speed network. In Britain all was not well. As far as transport was concerned, Morton likened southeast England to a river basin, "vigorously fed by strong flows of efficient and high-speed rail and motorway traffic from the north and west to the Thames at London and, after the Channel Tunnel is built, from Europe inbound to the Kent coast as Folkestone." If Kent was not to be faced with a huge motorway building programme, then the rail system must cope with much of that traffic. But he warned: "Unless such river basins have adequate drainage they turn into economically useless swamps – and that I believe is the prospect for the Thames-Channel region after 1995, unless we come up with a rail plan now." Morton concluded: "I say to the government, the fixed-link plan is not yet complete. Britain's road and rail system north and west of London is good to excellent; the Tunnel is going into place, and so is the initial road system between the Thames and the Channel. The last piece of the jigsaw is rail from Thames to Channel. Not a whole new line across Kent, but one or two 100-120 mph clearways, largely over existing routes – to be developed, I would say to the Treasury, by a partnership between private capital and British Rail."

Eurotunnel would be happy to become involved, he said. This was to become a favourite theme of Morton's – possibly only second to criticising the Channel Tunnel contractors – in the coming years. As the inaction on the British side of the Channel continued, so he became more and more frustrated, especially with what he saw as the Treasury's lack of understanding of the importance of infrastructure to the economy.

Almost six years after that speech, and eight years after the Channel Tunnel Bill was published which contained details of British Rail's planned international passenger station at Ashford, little or nothing has been done. According to the Department of Transport, no high-speed rail link between London and the Channel Tunnel is likely to be completed before the year 2000. And that looks very optimistic.

After the excitement of 1987, 1988 was a year of consolidation, of getting construction up to speed. With the UK construction industry buoyant, TML was struggling to attract sufficient staff and site workers of the right calibre and had embarked on a worldwide recruitment campaign. The construction boom had meant that builders were looking to source materials and workers from anywhere they could find them, and if that meant looking overseas, so be it.

In April 1988, Site Engineering Services, a firm supplying engineers to the construction industry, started recruiting in Australia in an attempt to fill the vacancies they had on their books. The response was overwhelming. Within days of placing advertisements in Australian newspapers, the firm had received 250 replies and had offered jobs to more than half of them. Some of these were expected to work on the Channel Tunnel.

Meanwhile, TML itself tried recruiting in Hong Kong, Singapore, Canada, Belgium and Holland. It also tried Cairo, much to the irritation of the British contractors working there, who accused the Channel Tunnel management of trying to poach their staff.

Scottish firm Lilley Construction was working on the huge Cairo Wastewater project, which also involved a large amount of tunnelling. Managing director Jim Barrowman was furious at advertisements in English in the local press inviting tunnellers to come to an interview at Cairo's Hilton Hotel. "The adverts are aimed at particular British contractors working in Cairo. The fact that they appeared in English in a local newspaper suggests they were not trying to recruit local engineers." He added wistfully: "There used to be some honour among tunnellers."

TML denied poaching but added: "We need 300 engineers and we are prepared to go around the world to get them....We don't feel that recruiting staff in Egypt or Hong Kong is any different from recruiting staff in the UK."

The problem had been made worse because one major source of good-quality engineers, America, had been ruled out because of the high wages they would command, TML said.

TML was taking on more people by the day – it started out employing six people early in 1986 but would have 15,000 on its books at the peak of the contract – and letting increasing numbers of subcontracts for the transport system as well as the civil engineering work. The pace of change meant old issues moved back on to the project's agenda. These were the nitty-gritty matters of construction, a world away from international high finance and its trappings.

Industrial relations had remained good to date, apart from the initial disputes at the Isle of Grain and Sangatte. But the British and French unions decided to ensure that if trouble did start, they would work together to avoid being played off against each other.

The policy began to take shape in 1987, when a meeting was called by the European Commission to discuss the Channel Tunnel. Those attending included Eurotunnel, TML, the railways and the British and French construction unions. This was followed up early in 1988, when the French union CGT requested a meeting with Ron Todd, the general secretary of the British Transport and General Workers Union. The French were rumoured to want to press TML to use the existing British terms and conditions in France. TGWU construction secretary George Henderson said he saw no reason why the unions should not have a joint-venture agreement when the employers also had one.

In June 1988 a further meeting was held in Brussels, at which the possibility of the British and French unions taking cross-border action in support of each other was considered. To date, industrial relations had been far worse at Sangatte than at Shakespeare Cliff, and the likelihood was the British would be taking action in support of the French. But by the end of the year it was the British construction workers who were threatening to bring the project to a grinding halt. Shop stewards on the British side were warning that relations at the Isle of Grain tunnel lining factory were becoming strained by poor pay and conditions. They especially stressed the pressure of work, the noise and the air pollution in the buildings. The Isle of Grain factory workers earned about half of that earned by their colleagues at the tunnel face.

Another background issue was the cost of some subcontracts. This affected Eurotunnel as much as TML. Under the procurement section of the contract, which mainly covered the locomotives and shuttle wagons, it simply paid TML the costs of certain items plus a management fee. Under the target cost section it paid the contractor the actual costs plus a fixed fee. If it comes in under budget, the contractor gets a bonus. If it is over budget, the contractor pays for 30 per cent of the excess or 6 per cent of the total target cost, whichever is less.

All either client or contractor could do was to stress to potential suppliers that as a truly international project it could shop anywhere in the world. This it did, with co-chairman André Bénard saying that Eurotunnel might be forced to look to countries in the Far East, such as Taiwan or South Korea, for help. With the threats made, little more could be done. But the problem of rising costs did not go away.

With TML slipping behind schedule early on, Eurotunnel decided to boost its own project implementation team, a kind of back-up group given the job of overseeing TML's work. It wasn't exactly a vote of confidence.

Details were announced in April, as Eurotunnel published its first annual report and accounts since the Equity Three share offer.

The two co-chairmen explained: "Our objective is to ensure that TML not only keeps to the planned programme but also puts the required management depth, certainty and coordination into the planning and design of the project – especially the transport system." This was a further sign that relations between Eurotunnel and TML were not in good health. And it was a clear sign of new trouble on the horizon.

Since Durand-Rival's attack on the contractors just before the Equity Three launch, construction rates had improved. The annual report said that work on the UK side was now "to schedule prior to a planned shutdown at Easter".

The French, meanwhile, were still three months behind, partly because of the financial collapse of one of the firms involved in building the tunnel boring machines. But they were optimistic about catching up. By the end of January 1988, the first French TBM had been delivered to site. It was ready to start work by the end of the following month.

This quick start-up was one of the advantages of the Sangatte access shaft, which allowed the 400 tonne main body of the TBM to be lowered in one piece to the opening of the tunnel. At Shakespeare Cliff, the machines had to be taken apart and reassembled underground.

To get back on schedule, the French had to bore 2 km by the end of the year. The UK TBM had to complete 5 km in what was expected to be by far the better ground. By mid-April 1988, the British had completed 1200 metres. Morton said: "While much remains to be done, we can be pleased that construction has made so much progress on both sides of the Channel." But he and Bénard were becoming increasingly critical of TML management and that criticism was to become more and more public in the months ahead.

Within weeks of Eurotunnel's annual report, its optimistic statements about construction progress looked hollow indeed. The British service tunnel TBM was stuck in an unexpected patch of sodden ground, described by one of the tunnel miners as "like wet pancake mix". The machine had barely moved since the Easter shutdown.

The French TBM would have coped. It had been designed for wet ground; the British Howden machine had not. The Howden TBM had begun its journey on the line of the failed 1970s project. The extensive site investigations had shown that the chalk was a good tunnel medium and quite dry. Accordingly, the Howden TBM was what was called an open-face standard rotary machine. It reached its starting point at the base of the cliff via access tunnels sunk from the top of the cliff in the early 1970s and again in 1986 and 1987.

Eurotunnel's technical director was one of the world's top tunnelling experts, Colin Kirkland. "The 1970s tunnel had stopped in excellent dry chalk and investigations had indicated that would be the case in the early phase of the work," Mr Kirkland later explained. "But it turned out to be a lot wetter and more fissured than that and the machine struggled in the first weeks."

Because the machine was designed to operate in dry, self-supporting chalk, its operation involved leaving a 1.5 metre length of newly excavated chalk unsupported before the concrete lining was fixed. If the chalk had been dry, that would not have mattered. But it was not. Instead the chalk turned out to be micro-fissured, allowing large amounts of water through. In the 1.5 metre stretch of tunnel where there was no lining, the water caused the chalk to cave in.

The only way round the problem was to try to modify the machine while it was underground, so that it could cope with the wet conditions. What the contractor would do if the modifications failed did not bear thinking about. One tunneller later said: "I thought the project might end, then and there."

TML had always planned that in poor ground it would line the tunnel with cast iron segments rather than concrete, as they were more watertight, but in these first few months it had exceeded the total amount of cast iron it had estimated for the whole project. The upshot was very slow progress on the British side, regularly below the 200 metres a week target and way down on the TBM's design speed of 5 metres an hour.

Meanwhile, in France, progress had also started slowly, but this was much as expected with the ground well-known for being wet and highly fissured. As a result, the tunnel boring machines were designed to operate in "closed mode" when the ground conditions were bad (as distinct from "open mode" used in good, dry ground).

Closed mode means that the machine is sealed off from the dirt around it and the face of the machine – the cutting head – keeps pressure on the dirt ahead as it churns it up and disposes of it. The machine can therefore work in the dry because the pressure inside it is slightly higher than that in the ground outside.

By the time of the service tunnel breakthrough in 1990, the reputation of the tunnelling teams was sky high. "What we have done here is revolutionise tunnelling," said Kirkland.

But in 1988, that reputation was anything but good. What mainly mattered as far as Bénard and Morton were concerned was whether the contractors were on time or not. And they weren't.

In the last week of August 1988, Eurotunnel formally notified TML that it was failing to meet the agreed project schedule. The move was taken under clause 46 of the contract, which required TML to submit a course of remedial action in writing.

The delay meant that TML was liable for substantial financial penalties. Two out of three "milestones" (target dates contained in the contract; failure to hit them could be penalised) had already been missed and Eurotunnel said that the fourth – 5 km of seaward tunnel dug from the British end by November 1 – was very unlikely to be met.

Details of the damages payable were not released by Eurotunnel, although Morton said they would be "well into six figures".

TML was upset, but Eurotunnel had also invoked clause 9 in the contract which forbade the construction consortium from talking to the press.

Again, what really bothered Eurotunnel was not just the construction delays but also how TML was being managed. This simply exacerbated the ill-will, which was now in danger of becoming as much a threat to the project as the unexpectedly wet ground off Shakespeare Cliff.

Construction News commented at the time: "Whether the contracting management falls short of the demand is not within our competence to judge. It would certainly be presumptuous to allege that the top end of Europe's contracting industries is not up to the job." And the newspaper warned: "One of the prime factors in timely completion must be unanimity of purpose between client and contractor. This seems to have been marred from the outset: public expressions of dissatisfaction with construction performance cannot help build the mutual confidence essential to success."

Towards the end of August, conditions slowly began to improve and tunnelling rates picked up. TML had always been confident they would. They had had to be. Through the bad ground the British service tunnel TBM had at times been managing just 60 metres a week. If that had continued for too long, the project would have collapsed. It would have taken years longer than planned to complete and the banks would almost certainly have withdrawn their support.

The British contractors were due to have completed 5 km of tunnel by November 1. With 11 weeks to go, they were still 2.7 km short of the mark and were progressing at best by 123 metres a week.To hit the target that rate would have to be doubled. They were doing better, but it was still not good enough, as Morton kept reminding them and the media.

Eurotunnel had good financial reasons for keeping the pressure on TML and for doing so in the full glare of publicity. Having spent most of the money raised in the various share issues, it was due to draw down the first £50 million tranche of its £5 billion bank loan early in November. It was clear the contractors were struggling, and the banks had to be convinced that Eurotunnel had the project under control if they were not to have second thoughts about their commitment to the project. But knowing that there was almost certainly more to Eurotunnel's agenda than criticising and embarrassing the contractors did little to appease them. They were forbidden from talking directly to the media, but they made their views well known to construction journalists.

What especially infuriated them was not so much the row itself – the contractors would hardly expect any client to be over the moon about such delays – but the manner in which they were being criticised.

At the end of August, Morton accused the British TML members of being more interested in making large profits out of the construction boom than in working on the Channel Tunnel, and he said there was not a strong enough commitment to the project among the top management of the contractors. TML and its member companies were apoplectic. As far as they were concerned they had encountered and had overcome a series of huge obstacles. Getting started would have been difficult enough in ideal circumstances, but here were ten contractors from two countries together trying to construct one of the biggest civil engineering projects ever attempted. Add to that the failure of one of the tunnel machine building firms; the financing and parliamentary delays; and the atrocious ground just off Shakespeare Cliff, and it is easy to see why they felt wronged.

"We have achieved a huge amount," said one TML middle manager. "All Morton is doing is making the job harder by lowering morale on site."

One of the most experienced construction industry commentators was John Allen, then editor of *Construction News*, a position he held for 28 years. What was happening, he said at the time, between TML and Eurotunnel would be instantly recognised by anyone with any experience of construction and tunnelling in particular. "It is quite simple: TML is suffering from a series of the kind of difficulties faced by all tunnelling engineers on all tunnelling projects." Deadlines are contractual risks, he said, and contractors would normally have a component in their contract for unforeseen difficulties. "But unforeseen difficulties do not translate comfortably into glossy brochures and publicity. Potential investors with little or no knowledge of the construction industry would not have responded favourably a year ago to perhapses and maybes. And what they were presented with was a bullish and, as it turns out, optimistic business plan."

Allen added: "It was a gamble and it has not paid off. Tunnelling – above all other civil engineering enterprises – very rarely turns out to be the operation it appears from the surface."

Allen also pointed out that if the bad ground conditions had not been picked up by the geological surveys, then TML would have grounds for asking Eurotunnel for extra time and money for sorting the problems out.

Such wrangling would be normal on contracts far less complex and far smaller than this one. "But these wranglings are and will be made extraordinary by the fact that they are being played out under microscopic examination in the full light of public attention."

Morton's aggression, he argued, was in danger of perverting the roles of client, engineer and contractor. The result was growing pressure on the engineer and the contractor to produce the promised goods in circumstances over which they had little or no real control. What the long-term result of that pressure would be was unclear, but there were fears that it could have some nasty side-effects if all the protagonists were not careful.

Allen mentioned one area that bothered him in particular – safety. There had already been two accidents. One had happened just before Christmas 1988, when four carriages on one of the construction trains broke their couplings and careered downhill to the bottom of one of the adits. In the second, a construction train hit and punctured a liquid petroleum gas canister. Luckily no-one was hurt in either incident, but the Health and Safety Executive prosecuted the five British members of TML in both cases. They were fined a total of £5,800 for the first accident and £20,000 for the second.

Tunnelling is always a risky business, and Allen was worried that the pressure to regain the time lost to date might make it more dangerous still. He commented: "There are fears that financial pressures and possibly damaging publicity could increase safety risks underground if tunnelling is accelerated too quickly in order to stem criticism."

Both Eurotunnel and TML said that safety would not be sacrificed for speed, and the HSE insisted that they would continue to ensure that safety was not compromised.

When Eurotunnel presented its interim results and report on October 3 1988, the cash cost of the delays became clearer.

There had already been stories in the press speculating that Eurotunnel might have to launch another rights issue, or that the banks behind the £5 billion loan might demand tougher conditions in light of the delays and the rows with TML. Eurotunnel now insisted that the loan was not under threat at all, and said that the details of the loan would be finalised in time for the first draw-down in November. Eurotunnel also said that since August the delays had been cut from six months to three on the French side, and from three months to six weeks on the British. But it added that its estimate for the final cost of the Channel Tunnel had now risen by 7 per cent, to £5.23 billion from the prospectus figure of £4.87 billion (both figures were at July 1987 prices).

Not all of the £360 million increase was down to rising costs. It included the cost of improvements to the terminals, for example. But the report said that much of it was accounted for by the cost of recovering delays in the tunnelling work, projected cost overruns and by Eurotunnel's decision to set up a large project management team of its own to oversee the contractor.

This was to prove a source of increasing friction between client and contractor, who saw Eurotunnel's Project Implementation Division, as it came to be called, as duplicating their own and the MdO's efforts and hence slowing down decision making and increasing bureaucracy.

The PID was certainly not a small commitment by Eurotunnel. Morton said that Eurotunnel had taken on more than 100 extra project management staff in the past six months, bringing the total employed in the division to more than 300. (Significantly, their numbers included 35 on secondment from the giant American contractor Bechtel, a firm which, although neither British nor French, would have an increasing influence on the tunnel as time passed.)

Although Morton suggested that some of the extra people had been recruited to work on the transport systems associated with the project – which were proving trickier than predicted – there was little doubt that they would also have the job of keeping close tabs on TML.

Eurotunnel's technical director Colin Kirkland said the PID had played a key role in solving the initial tunnelling problems. Asked if Eurotunnel had created a powerful project management team of its own because of TML's poor performance so far, he said: "I wouldn't quibble with that."

The 7 per cent cost increase had little impact on Eurotunnel's estimate of its long-term profitability as at the same time it announced increased traffic forecasts which it said would boost revenues in 1993/94 by 6 per cent, largely because of strong growth in the world and especially the British economy since 1985 – the base year for the prospectus forecast.

As predicted, TML did miss its November 1 deadline. By then the service tunnel TBM should have been 5 km from the English coast towards France. Just 3.6 km had been completed. The good news, though, was that tunnelling rates were still rising, and sometimes now reached 170 metres a week.

The size of the financial penalty incurred by TML was not disclosed, although reports at the time said it would increase every day until the 5 km milestone was reached, up to a maximum of £10 million by the end of January.

The City of London was untroubled by all the publicity. By early December the shares had soared to 465p. At the end of 1987 they had been offered at 350p but had plunged to 250p after the first day's trading.

On site, relations between TML and Eurotunnel staff were reasonable. But at board level the bitterness and wrangling continued. In December 1988 the French members of TML called a press conference in Paris to put their side of the story. They were incensed at the way they had been treated by a client they and their British colleagues had created. The five company chairmen could not speak as TML, as this would be in breach of their contract, so they said they were speaking on behalf of their individual firms.

For a contractor in dispute with a client to give such a press conference was virtually unprecedented. When problems arise on a job, the reasons are usually far more complex than incompetence or bad judgement. At least part of the blame is often due to the client or the design team – perhaps information given to the contractor was incorrect or was changed at the last minute – but even in such circumstances the contractor very rarely makes a public song-and-dance. There may well be a dispute and it may well go to law, but construction is a service industry and the contractor might well want to work for the same client again in the future, so public embarrassment or worse is out of the question.

The five firms were Bouygues, which claimed to be the biggest construction company in the world, Spie Batignolles, SAE, SGE and Dumez. They had, they stressed, top quality credentials, huge experience in tunnelling and, what is more, they had seconded some of their best people onto the Channel Tunnel project. And they were optimistic of clawing back the time lost to date.

Francis Bouygues, head of the Bouygues empire, pointed out that these companies were among the original promoters of the scheme and, as such, were just as aware as Eurotunnel of the problems and the need to overcome them to make the project viable. "We are not just contractors," he said. And he added that some of the group regretted the decision taken in 1986 to create the client and abandon the role of promoter.

However, the overriding feeling from the contractors was one of disappointment rather than anger. The Channel Tunnel should be one of the great civil engineering projects of the century. It should be executed with dignity and professionalism. Yet here were the main players sniping at each other through the media.

During the Paris press conference, Dumez chairman Jean-Paul Parayre, one of the key figures in winning the mandate for the Channel Tunnel Group and France-Manche, stressed that TML would be disputing Eurotunnel's claims and would be asking for compensation under the contract. He refused to say more, but behind the scenes the contractors were taking concrete action to defend themselves and their reputations. Eurotunnel was left in no doubt that they thought they had a very good case.

In fact, the two sides had been negotiating since June. (Eurotunnel later described it as an "arduous confrontation" with TML and its shareholders.)

The talks were to last nine months, but in December 1988 there was no end in sight. The core issue for Eurotunnel was how to keep the project on the track outlined in the November 1987 prospectus, and to that end it wanted to see management changes at the top of the contracting organisation. In response, TML was claiming extra payments and time to cover the unforeseen tunnelling problems and what it argued were a number of late changes in specification. It also presented a series of claims based around the late ratification of the Anglo-French Channel Tunnel treaty.

The problem here was that the late approval of the legislation had in turn delayed the fund-raising process which meant that TML was forced to delay starting on site.

In total, there were 40 cash claims totalling £47 million and time claims of 11 months.

The relationship between the two sides was now coming to the boil. Morton was still insisting on management changes at TML. The contractor was insisting that the delays were largely out of its control and was demanding compensation. In such an atmosphere, rumours were flourishing. A number of Japanese construction firms were reported to have offered their services to Eurotunnel. A number of leading contractors were supposed to want Morton silenced or moved from his post to a less "sensitive" position. But as the project seemed to be lurching towards a boardroom crisis, under the Channel it was going better than ever. An explosive 1989 was in prospect.

11 All change at TML

"The moment you have broken through, the tunnel has lost all its magic. It becomes just a hole in the ground connecting two shafts."

David Denman, Senior Agent, TML

The year 1989 could not have started worse for Transmanche Link.

Just as tunnelling rates were starting to pick up and put some of the disappointing progress of 1988 behind the contractor, tragedy suddenly struck the site.

Andrew McKenna, a 19-year-old surveyor's assistant, became the first man to die on the project, crushed between two trains in the service tunnel.

The site team's morale dived at the news, and union chiefs and politicians questioned whether TML's drive to increase the rate of tunnelling was being carried out at the expense of site safety. And when David Simes became the second site fatality just two weeks later, there were clamours for a full safety review of the project.

These were desperate times for everyone touched by the deaths – TML, the workforce and the family and friends of McKenna and Simes. With two men dead in two weeks, TML was starting to get an idea of how difficult it would be to protect its workforce.

Safety systems had been put in place on the project but still accidents continued to happen.

TML did not handle the publicity around the accidents well, and were widely condemned for a deafening silence over the incidents. But the management believed it was truly striving to make the tunnel as safe as possible. But something was going wrong, and whatever it was had to be isolated.

It would actually take the contractor more than 18 months and three major project audits to come to grips with the level of safety required on site. These were 18 months in which seven men would die on the tunnel on the UK side alone, and another two in France.

TML's site problems at the beginning of 1989 were not made any easier by the turmoil surrounding its management structure. At the time of the deaths of Simes and McKenna, this structure was about to be totally overhauled. And despite the faster pace of tunnelling, there were still deep divisions between the contractor and its client Eurotunnel.

The TML chiefs found it difficult to forgive Eurotunnel co-chairman Alastair Morton for his accusation the previous year that they were responsible for the delays in tunnelling. TML was forbidden under the contract from speaking about the project in public, so rendering the war of words being conducted by Morton somewhat one-sided.

The row eventually gave rise to the rather bizarre spectacle of the chiefs of the five French member contractors of TML calling a press conference in Paris in December 1988 to show off their talents as contractors to 200 members of the world's press – an amazingly high number for a briefing on an industry not previously seen as sexy enough for good copy. The Channel Tunnel was at least changing that perception. The whole event was somewhat bizarre, one that British contractors would be naturally reluctant to set up themselves, but to the French it was a matter of honour to defend their good name, which was under attack from Morton.

The Paris show was the first inkling of the strange events that were to come to dog the project. And it was not the last time that the contractors held Paris side-shows during the construction.

Despite, or perhaps because of, the press conference the situation at management level was verging on farcical. Delays on site were costing money and Eurotunnel was not keen to pay. A penalty clause in the contract allowed Eurotunnel to fine the contractor for missing progress milestones – this it did three times, to the tune of around £10 million. But as relations between the two sides deteriorated, the costs of the project continued to rise and the delays continued to mount up, with tunnelling now up to six months late. A deal needed to be struck, and quickly.

The feeling on site was that the two sides were not working well together at management level, and this was causing frustration at every level. Changes needed to be made if the project was to get back on schedule and if the running dispute between the two sides over payment for tunnelling difficulties was to be settled.

But Morton's unhelpful remarks about TML's management seemed to be hindering any restructuring of staff rather than helping. TML did not want to be seen to be bowing to the "bullying" demands of its client, even if the contractor itself recognised the need for change.

The first move came at TML at the beginning of February 1989 – and it was a dramatic one. TML chairman and chief executive Andrew McDowall was demoted to deputy chairman and Phillippe Essig, who worked for the French railway operator, SNCF, took his place as chairman. The move was clearly a victory for Alastair Morton and Eurotunnel, who had been becoming increasingly frustrated with the TML management, particularly as site progress was still so slow. Relations between Morton and McDowall had become strained, with Eurotunnel feeling that McDowall was elusive and had more on his plate than just the Channel Tunnel scheme – something McDowall himself strongly denies. "The TML management was totally independent of the parent companies. In fact we were criticised for that very reason."

He added: "I have a lot of respect for Morton, but we did not see eye to eye because of what I regarded as his interference in the contract."

By now it was becoming increasingly clear that what Eurotunnel actually wanted in terms of performance from its contractor – and the sort of project it was expecting to receive in 1993 – was not quite what TML had expected to have to deliver. It did not take much analysis by the contractor's team to recognise that multi-million-pound disputes were hiding just around the corner.

Both sides had also recognised that the contract was too vague to be a clear arbiter in any dispute. So each row was destined to run and run. The lawyers must have been rubbing their hands together in glee.

In short, TML needed a full-time team to fight Eurotunnel for every penny on the project.

Eurotunnel welcomed the personnel changes because it also wanted to see a fresh management team in place, dedicated to the single task of completing the Channel Tunnel. And it was making appointments itself. Dr Tony Ridley, a former chairman of London Underground, was appointed joint managing director with Alain Bertrand from SNCF. Further changes were clearly on the way, with the contractor promising to appoint a new chief executive and a technical director responsible for construction as soon as possible, in a sweeping change to its structure.

On the basis of the executed and proposed changes at the top of TML, Eurotunnel was prepared to do a deal with TML on the claims for the tunnelling delays. The two sides had been quietly discussing the thorny question of who should pay for the hold-ups for almost six months. Because TML had missed three tunnelling milestones – laid down in the original contract – Eurotunnel was able to penalise them financially. The contractor had argued that the slow progress had been caused by unforeseen ground conditions and financing difficulties at the beginning of the project. And it was arguing for an 11-month extension of time that was expected to cost the project in the region of £500 million.

That a deal was made in spring 1989 was due to the management changes under way at TML – a fact that Eurotunnel highlighted in its 1988/89 annual report.

The deal was so successful that at first it seemed likely that despite the poor progress TML would still be capable of hitting the target opening date of May 15 1993. The contractor shifted its position from demanding an 11-month extension of time down to a more reasonable three months. This would take the official completion date to August 1993. TML would then pull out all the stops and bring the tunnel in to the original schedule – May 15 – but be paid a bonus for doing so. This is not an uncommon tactic in construction projects and acts as a way of compensating a contractor for time lost which may not have been its fault, while maintaining the original programme for the concerned client. But after detailed discussions both sides accepted that the delays were already too great to allow adhesion to the original programme.

When the details of the deal – called the Joint Accord – were released by Morton and Bénard in April 1989, they had to admit that the tunnel would now open one month late, on June 15 1993. It was the first slippage in the programme and a disappointment for Eurotunnel, which was determined to get trains running across the Channel as early as possible. Despite this the mood was positive. Morton described the negotiations with TML as a year of "arduous confrontation" resolved with a "tremendous improvement" in relations between the two sides.

The deal was set up to give TML an extra £106 million in incentives and cash, previously withheld by Eurotunnel, if it hit a series of new, and now more realistic, milestones allied to the new programme. The delay and the deal put the estimated cost of the project up to £5.45 billion from the £5.227 billion estimate six months earlier. This compared with an original estimate for the cost of the tunnel of £4.877 billion.

Although Eurotunnel had access to £6 billion in funds before it had to seek more money, it announced that it was planning discussions with TML on how to cut costs to save some of the £223 million which the deal was costing the client. Although most of these cuts were to fall on the TML staff on the British side, the atmosphere between the two in April 1989 seemed uncommonly good.

With a new deal under their belt there seemed every reason for believing that the work could be completed in relative harmony.

Phillippe Essig was in place for TML and Tony Ridley for Eurotunnel. Both had railway backgrounds and knew each other from old, so seemed quite capable of working together in a "new spirit and structure of communication between the two organisations", according to Eurotunnel's latest annual report. Unfortunately that spirit was not destined to survive very long.

The Eurotunnel annual report welcomed Essig but was not shy in encouraging the contractor to hurry up other management appointments. "The early arrivals of a new chief executive and a technical director to support Phillippe Essig as new chairman of TML are essential to maintain momentum and indeed to the Joint Accord", it said. The contractor still viewed such public expressions of dissatisfaction with its team as provocative in the extreme, and with some justification. The two sides met privately often enough for any grievances or obstacles to progress to be discussed fully. And TML felt that it had spent much of the early part of 1989 satisfying its demanding client by replacing senior managers – a development hindered, not helped, by every public attack on TML's ability.

In some ways the wounds inflicted at this stage of the project could never heal. Although TML was happy with its part of the deal carved out in the Joint Accord, there had been an element of public humiliation piled on 10 of the biggest civil engineering contractors in Europe. Those contractors were not about to forget that. They felt that for all the boardroom difficulties they were still doing a good job on site, where it mattered.

The central service tunnel, which the British and French had been carving out during the past 18 months, was starting to show real signs of progress. The two tunnelling teams were regularly converging at more than 400 metres per week, more than four times faster than most weeks in the previous year. The stretch of poor fissured clay on the British side was well behind the service tunnelling team, and the site transportation system was beginning to operate to such a level of efficiency that it was more than capable of keeping up, even with the improved rate of tunnel boring.

The service tunnel boring machine had been modified underground to take account of the wet ground, and this operation, difficult as it was, seemed to have been a success. On the French side, although the ground was presenting predictable difficulties, progress was improving as the teams got used to their work. So much so that the project was about to get its first major boost – a tunnel breakthrough on site.

For a tunneller a breakthrough ceremony is a strange experience.

After months, maybe even years, battling away underground with construction colleagues to carve out a hole in the ground, suddenly all and sundry are in the tunnel helping to finish the work off. Having done the difficult bit – faced the dangers of flood, collapse and crushing by machinery – the miners are suddenly joined on site by the management, quickly followed by the media. Now it is safe down below men and women appear from nowhere, wearing brand new wellies and shiny hard hats which stand too high on their heads because the straps have not been properly adjusted. They get to stand in a tunnel specially cleaned for their visit. Everything is tidied and cleared. God forbid that the guests should have to clamber over wagons and tools, ducking below lighting strings and balancing on impossibly slippery railway tracks! Instead they are given a cordoned-off area in which to stand to watch the tunnel boring machine break through with a great roar and mushing of muck.

Of course, the assembled guests rarely see a real tunnel breakthrough. That has happened several hours earlier – achieved by miners at 3.20 a.m. when the world's press is sleeping off the hospitality at the local hotel.

What the press sees is a plug of concrete or earth – often ludicrously painted with the contractor's and client's logos – removed just hours after it has been put in place by the same teams who broke through properly earlier that day. The tunnel boring machine has been waiting immobile behind the plug until everyone is assembled, after coffee and biscuits and just before the champagne reception. So for the tunnellers the whole thing is rather strange.

They are, of course, invited to the party, where they are all very well mannered but look somewhat uncomfortable during the inevitable delays and photo-calls.

The traditional waving of hard hats in the air is normally suffered with dignity by the miners, but they don't really relax until the strangers have left the site. This leaves the way open for the best bit about a breakthrough – the celebration. But once that is over then a miner will explain that the tunnel – for months having held an air of mystery for the miners because the path ahead is uncharted – is of little further interest.

David Denman, a senior agent on the Channel Tunnel, summed it up: "The moment you have broken through, the tunnel has lost all its magic. It becomes just a hole in the ground connecting two shafts."

Of course, to the management of TML and Eurotunnel a breakthrough was a godsend. The first on the project came in April 1989, generating much needed positive press coverage and boosting morale, not only on site and in the boardroom but in France and Britain as well. The completed tunnel was the French landward service tunnel drive. It was just 3.25 km long and had taken ten months to excavate.

Although there are only three tunnels built under the sea, the two running tunnels and the service tunnel between them, in all there were 12 tunnel drives on the project. This is because driving was carried out from both sides of the sea, of course, but also because it is rarely possible to begin tunnelling from the very beginning of a proposed excavation. Tunnelling begins where it is most convenient to set up a tunnelling site, with all the back-up and support systems that such a site demands. As this is normally at some point along the proposed tunnel line and not at one end, tunnelling has to be carried out in two directions from the starting point.

In the UK tunnelling began at Shakespeare Cliff, right on the coast but nearly 8 km from where the trains will finally emerge into the terminal at Folkestone. So three tunnel drives started from that point seaward, eventually to join their French colleagues in a breakthrough under the sea, and three began to bore the 8 km landward to break through at the portal just short of the terminal.

In France the tunnelling began at the base of a 75 metre deep, 55 metre diameter shaft sunk near the coast at Sangatte, 3 km north of the terminal at Coquelles. Again, three drives started digging the tunnel seaward from the bottom of the shaft to meet the British miners, and three were needed to form the 3 km-long tunnels from the shaft to the terminal.

So on each side there was a seaward service tunnel drive, a landward service tunnel drive, a seaward running tunnel (north) drive, a landward running tunnel (north) drive, a seaward running tunnel (south) drive and a landward running tunnel (south) drive.

To confuse the issue further, only 11 tunnel boring machines were used for the 12 drives, because one machine did two tunnels – the two French landward running tunnel drives.

All in all the arrangement gave rise to nine breakthroughs on the project, three at each terminal and three underground, and so there were several parties. But there was something special for the teams about the first one. The French machines rather quaintly had names, and it was "Virginie" who appeared at Coquelles in April.

With the Joint Accord signed and the first tunnel excavated, TML was able to bring its good news in threes that May by announcing the appointment of a new chief executive. Jack Lemley, a 54-year-old American, was appointed for his extensive experience of tunnelling in senior positions for top American construction firms Blount and Morrison-Knudsen. TML's management restructuring was still under way, and just before Mr Lemley's appointment French director François Jolivet had resigned. Lemley was seen as a construction heavyweight and a man with a reputation for bringing in construction projects early and within budget, something of a tall order for the Channel Tunnel. His brief was to take the project through construction to start-up and full operation of the fixed link. He was immediately diplomatic enough to express his wishes that the team that had developed the project should stay in place, while at the same time making it clear that he would not be surprised if they moved back to their parent companies.

Although some expected a huge wave of management moves at TML, the mass exodus never really occurred. However some key changes did follow. In June 1989 John Reeve resigned from TML as British director general in a move widely predicted at the time, but also one seen as another victory for Eurotunnel's Alastair Morton, who was said to dislike Reeve's perceived tendency to be claims-orientated on the project. Reeve, who had also resigned from Costain where he had been a director for 14 years, was involved in TML from the start and did not want to leave. But the appointment of a technical director appeared to overlap his position and make it seem to him untenable. His resignation more than any other signalled a change in TML's structure. But if Morton thought he would have an easier time from the team coming in, then he was in for a shock.

The new technical director was Klaas van der Lee, a divisional director of TML member company Taylor Woodrow and, perhaps crucially, project manager of the Channel Tunnel during the 1975 attempt. Also coming on board was Jacques Thibonnier, as director of engineering and transportation.

If the new team expected a period of calm to ease them into place, then that was shattered by startling new problems for Eurotunnel on the costs front. Just as some of the fears were fading about escalating tunnelling costs, so Eurotunnel was hit amidships by the news that the cost of building the shuttle trains had rocketed. This news was broken by TML to its client just as Lemley joined the team. It was part of TML's brief to "procure" the contracts to build the tunnel's rolling stock. This it would do for a fee, with Eurotunnel paying the price of the work.

In 1986 Eurotunnel had told Parliament that it estimated that the cost of building the 522 shuttle wagons – which would take cars, coaches and heavy goods vehicles back and forth between the terminals through the tunnel – would be £160 million. By its 1987 prospectus Eurotunnel had revised its forecast to £226 million. When the contracts were awarded in July 1989 they totalled £600 million. Changes to the design of the shuttles to improve safety features had increased the cost dramatically, and Eurotunnel announced that the high price reflected the high quality of construction required for the tunnel.

Eurotunnel also announced that it was talking to its banks about raising more finance for the project. Already increasing interest charges and construction costs following the tunnelling delays had taken the expected outturn cost of the project dangerously close to the £6 billion of funds raised by Eurotunnel. But the increased shuttle bill had now forced the project cost to break the £6 billion mark, and news soon emerged that Eurotunnel was sweating over new construction cost estimates, fuelling speculation that prices were soaring out of control.

Morton stepped in to try to calm any City panic by describing the situation as "a routine summer crisis", and expressing confidence that the funding banks would continue their support. But he did admit that it looked likely that Eurotunnel would not have enough money to take it to the 1993 opening, and that it would run out if more money was not raised.

At the same time as the shuttle crisis, TML and Eurotunnel were engaged in agreeing a cost to completion estimate, a figure that the banks insisted was produced regularly. Unfortunately the two sides were making little progress in agreeing this figure, a fact becoming increasingly apparent to the media, the City, the shareholders and the funding banks. Eurotunnel would have to come clean on cost forecasts in its October interim report, and as the date approached speculation about the likely figures approached hysteria level, while Eurotunnel's share price plunged to around 550p, less than half its peak, reached in the calmer days of summer 1988.

When Eurotunnel did release the new estimate it had to admit that TML still disagreed with its calculations. Morton said that the estimate for the

EVENING STANDARD, 19 FEBRUARY, 1990

"Do you know Wilson, I haven't seen a boardroom battle like that in years!"

outturn cost of the scheme had risen to £7 billion, a 40 per cent increase from the original £4.8 billion. But TML, for its part, estimated that the project would now cost nearer £7.5 billion. Until the two sides could come to some agreement, Eurotunnel could not go to its banks to try to raise the extra money it required for the project. This was now between £1 and £1.5 billion, depending on the agreed project cost estimate.

Morton spent much of the October interim report press conference trying to convince everyone that "there is no brawl going on" over costs between TML and Eurotunnel; rather, there was just a "strong difference of commercial opinion." He described newspaper reports that there was a serious rift between the two sides as coming from the "Batman school of journalism". But he could not resist some public sparring and proceeded to blame the cost increases on TML, and condemned the contractor's claim that the terminals and tunnel fixed equipment would cost an extra £382 million as "manifestly absurd". Eurotunnel's figure for the work was £1.488 billion – TML's was £1.87 billion. This gap was the main reason that the two sides had different estimates for the cost of the scheme.

It was poignant that even at this early stage of tunnel construction the centre of the costs dispute was not the tunnelling, as expected, but the bill for the lump sum work.

This was the cost of building the terminals and installing the fixed equipment – the mechanical and electrical systems – into the tunnel. Up until that point much of the attention had been on the tunnelling delays and their consequence on the costs of the project. But what was becoming increasingly clear was that design changes to the system, and safety pressures applied following the lessons learnt from the King's Cross fire the year before were starting to impose a substantial cost burden on the project. The lump sum work would become the basis of the most bitter and long-running dispute on the tunnel, and Morton's condemnation of TML's claim was the first hint of the problems to come.

What was also of note in the October 1989 progress report was that TML and Eurotunnel did actually agree on one thing – that an estimate of the cost of the project from the banking syndicate's technical adviser was wrong. The technical adviser, employed by the banks to give an independent view, was estimating the final cost at nearer £8 billion. And the reason for this was that it believed the project would not be complete until December 1993. Both sides rejected this and agreed that they were determined to hit the June 1993 deadline.

But the row had to be resolved, and TML and Eurotunnel agreed that a third party should look at the project. The Channel Tunnel scheme was guided throughout by a Maître d'Oeuvre (MdO) – a French-style project manager which comprised a consortium of UK engineer W S Atkins and French consultant SETEC. It was agreed that the MdO would carry out an assessment of the costs and claims situation and report back in December 1989 in an attempt to solve the costs disputes and satisfy the banks that the cost to completion was agreed at an acceptable level.

Eurotunnel, TML and its engineers also began looking for ways of cutting costs, concentrating on any areas of over-design or duplication of systems or staff.

TML immediately let it be known that it believed savings could be made in Eurotunnel's Project Implementation Division, which had been set up to monitor TML's progress. At first the PID comprised a relatively small team of construction professionals, mostly seconded from the American construction giant Bechtel, a firm with something of a reputation for construction troubleshooting. But the division grew massively until it had some 350 staff, and was known to TML as the "shadow contractor".

TML claimed the PID duplicated much of its work and was actually creating work and extra cost because TML had to respond to the mass of correspondence emerging from the department. TML's demands that it be scaled down were heeded by Eurotunnel, and the staff numbers were reduced from 350 to 270. TML's management was hugely boosted by this success, which went some way to balancing their agreement to make their own management changes under pressure from Eurotunnel earlier in the year.

Fortified by this success, the new TML management team looked for other changes – particularly the removal or demotion of Alastair Morton.

Morton had seriously offended the contractor with his attacks on tunnelling progress, which were widely seen as unhelpful and ill informed. TML managers felt that Morton could not resist a dig at them at any opportunity, and as a result relations between the two sides had become so sour that TML was forced to change its management. This still rankled with the contractor, and TML's chiefs wanted Morton to suffer the same fate as the former TML managers. They began to work towards getting rid of the Eurotunnel chief.

The end of 1989 became a period of high drama. TML took the opportunity of the potential funding crisis to propose new deals to Eurotunnel aimed at cutting costs. But these deals included demands about Eurotunnel's management – involving the distancing of Morton from the project – and were deemed unacceptable by the developer.

The banks were growing increasingly impatient with the inability of both sides to agree how much the project was going to cost, while all the time both sides became more entrenched in their positions. In December 1989 Morton told the news agency Reuters that "unless we resolve our negotiation with the contractors by early January the banks will get fed up with financing the project." But still the row continued.

Any question of Eurotunnel and Morton budging from their stance on the fixed equipment costs seemed to be put to rest by the Atkins-SETEC report. This horrified TML by backing Eurotunnel's cost estimate of £1.488 billion for the fixed equipment work.

But instead of resolving the dispute the report just aggravated it. It was not a binding contractual ruling but an independent estimate of how much the fixed equipment work – which had not yet been carried out – was going to cost.

Eurotunnel hailed the report, claiming that it vindicated its case totally. TML politely accepted the independence of the report but said its cost estimate was "rock bottom". TML also pointed to the MdO's assessment that the contractor did have a case for some extra payment because of design changes to the fixed equipment, which was TML's argument all along. TML just argued that the MdO's assessment of that cost increase was too low.

Not for the first, or indeed the last time did a vital document on the project manage to be something like a pronouncement from the Delphic Oracle – open to interpretation in favour of both sides.

As the New Year began the two had signed no deal and Eurotunnel was running out of money. The banks were not willing to open lines of credit unless there was a deal, and the financial pressure on Eurotunnel was mounting.

On the first Monday of the New Year TML announced it was taking Eurotunnel to court in France because it had not been paid its monthly account. Suddenly the greatest civil engineering project in the world was on its knees. The banks were nervous, Eurotunnel had little cash and the contractor was not being paid and was dragging its client into court. In an industry used to project disputes it takes a lot even to raise an eyebrow, but now the world watched in astonishment. Rumour and counter rumour were reported by a media that had suddenly found that construction could be fascinating after all. Calls to have TML thrown off the site and replaced by Koreans, Japanese and any other contractor that was not British or French were followed by tales of secret deals between the funding banks and TML to have the contractor take over the project and Eurotunnel removed.

Whether it was TML's legal action, luck, diplomacy or just sheer panic that allowed TML and Eurotunnel to forge a deal on costs is not clear. But with the whole operation falling down around their ears, and just 24 hours before a meeting of the funding banks which might have brought work to a halt, the chiefs on both sides made sufficient progress towards a deal to convince the banks to unblock the lines of credit and allow the drawdown of £390 million – enough for about four months' work. But, the banks said, by then a proper agreement between the two sides on the project out-turn cost must be in place.

This deal was a rather complex affair, and it clearly dodged key issues which still needed sorting out. It was a new agreement on the price of the tunnelling work.

The original contract said that Eurotunnel would pay £1.25 billion for the tunnelling. It would then pay 70 per cent of any increase in costs, with TML paying the other 30 per cent. However, TML had a cap on its contribution of 6 per cent of the total cost of the tunnelling. What this meant in reality was that once cost overruns hit more than £312 million, TML's contribution to extra costs reduced dramatically. And the estimates showed that the magical £312 million would indeed be passed.

In the new deal, Eurotunnel, bearing in mind TML's claims for poor ground conditions, agreed to pay the first £1.58 billion of tunnelling costs, with TML paying 30 per cent of costs over that. But this time, crucially, there would be no upper limit.

The procurement element of the contract – which covers the purchase of the shuttle trains and rolling stock – had gone through the roof already, more than trebling to £600 million. This could not be blamed directly on TML. Indeed, as the contractor was on a percentage fee it would make more money because of the increased cost. In a spirit of goodwill TML agreed to reduce its fee.

The gaping hole in the deal was the lump sum work – where the row had been centred for most of the year. In lump sum work all the cost increases are borne by the contractor unless they can be shown to have arisen from changes to the work ordered by the client. This is the traditional way of contracting in the UK construction industry – and a dangerous one at that. This lump sum included the terminals and fixed equipment. TML estimated that the cost of this would be £1.87 billion – Eurotunnel put the figure at £1.488 billion. TML claimed the extra £382 million was due to changes to the design and specification of the project by Eurotunnel. Eurotunnel disagreed.

They resolved to go to arbitration on these costs, an agreement which was sufficient to woo the banks back to the project and get a promise of interim funding. Perhaps more importantly, though, Eurotunnel agreed to reduce the scope of this section of the work, not least by bringing down the maximum speed of the trains in the tunnel to 130 kph. This has the effect of dramatically reducing the heat generated in the tunnel and therefore allowed for a downgrading of the hugely expensive ventilation system.

It seemed a fair deal. It was by no means comprehensive but seemed to go some way to convincing the world that TML and Eurotunnel could sit down and make progress. So it was perhaps typical of this stage of the project that even in striking this deal, TML and Eurotunnel were able to create a whole new and quite remarkable row.

The deal was announced by Alastair Morton on January 11 1990, and by his standards the language used was calm and inoffensive. Certainly nothing seemed amiss, and the press reported the situation without recognising any attacks on the contractor. But perhaps crucially the announcement was not a joint affair between TML and Eurotunnel. Morton had rejected a proposal put forward by TML's chief negotiator, Peter Costain – group chief executive of Costain, a member company of TML – to issue a joint communiqué on the deal. It seemed a reasonable request, unreasonably denied. And the result, once Costain heard the statement, was more bad blood.

In a letter from Costain to Morton on January 16, Costain condemned the press release and subsequent briefing by Morton as "inaccurate, incomplete and calculated to mislead." He then listed a series of particular points that had riled TML.

Eurotunnel had "failed to make it clear" that this was the second round of cost reduction, and TML's target was a reduction of at least £200 million. "Eurotunnel's acceptance of this extra cost-saving commitment was an essential precondition for TML's signature of the agreement," he wrote.

Costain went on to deny that TML had accepted a maximum train speed of 130 kph – "this is still too high" – and said that TML never agreed to comply with the MdO's December 1989 cost review, despite statements to the contrary in Eurotunnel's briefing note. He then moved on to remarks that Morton had made about the UK side of construction. Once again, Morton had been unable to resist having a dig at the contractor and the briefing note had unfavourably compared productivity on the British side of the tunnel with that in France. Costain wrote: "I cannot understand what you hope to gain by the disparaging remarks about the performance of UK construction. Your comments do nothing to improve the relationship between us, they have an intensely demotivating effect on our employees."

The letter was sent to the four main, or agent, banks who led the 220-strong banking syndicate which was funding the project. It was the most tangible evidence to date of how much TML was enraged by Morton and how much they wanted him out.

Indeed, the letter touched on the thorny issue of staffing when it called Morton's reference to "trimming the project supervision" as "disingenuous". Costain wrote: "You are quite well aware that we would not have signed the agreement were it not clear that far-reaching senior management changes in Eurotunnel were irrevocably committed." And these, as far as TML was concerned, meant moving or sacking Morton.

Any good that might have been done by the announcement of the "mini deal" was destroyed by the briefing note and Costain's letter. The depth of animosity between the two sides was now so obvious that it seemed inevitable that heads must roll if the project was not to grind to a halt in embarrassing chaos. The contractor now openly – or as openly as it could within the restrictions of the contractual gag on public statements – called for a new post to be created at Eurotunnel. TML wanted one person at senior level to take over the management of the project, distance Morton and Bénard from the work and, in the words of one site source, "stop Eurotunnel's day to day meddling."

What is more, the contractor still believed it had a powerful weapon to wield. TML had not yet signed the January 1990 agreement, details of which were still being drawn up by the lawyers. Without TML's signature the banks would still not release the money to Eurotunnel, although they had agreed to do so once a deal was signed.

This tactic had a down side of course. TML was still not getting paid for its work, and indeed Eurotunnel could not afford to pay its contractor without the money from the banks.

TML was initially encouraged to hear that management changes were indeed afoot at Eurotunnel. When they saw the changes, they hit the roof. Morton had been moved from co-chairman to deputy chairman, but had taken the role of chief executive. The TML chiefs were apoplectic.

Although Tony Ridley was given the job of day to day running of the project, TML saw Morton's move as deliberately provocative. They wanted him removed completely, but would have settled for a sideways shift to keep him away from its work. What they seemed to have got was a management reshuffle which left Morton in a stronger position than before. The contractor made it clear that there was no way that the January deal would be signed until something was done about Morton.

As far as TML was concerned, the deal was conditional on management changes at Eurotunnel and they were not getting the changes they wanted, so they were not going to sign. At the same time, TML reinstituted proceedings in the French courts against Eurotunnel in an attempt to get paid for work already done on site. Payment had dried up totally since the turn of the year and the financial pressure was beginning to tell. Eurotunnel, for its part, made it clear that there was no point in going to court because it just did not have the money to pay.

Incredibly the row raged on, and for the second time in a month there seemed a very real chance that the project would fold and construction grind to a halt.

Nobody seemed to be interested in what was now going on under the sea, with breakthrough of the service tunnel just 10 months away.

The Channel Tunnel had become a multi-million-pound boardroom farce that was keeping the industrial editors of the British broadsheets and French quality press in full-time employment. French co-chairman André Bénard said what everyone was thinking when he confirmed that the row was so serious that the tunnel might not actually be completed. He expressed grave disappointment that such a great project could be brought to its knees by a clash of personalities, and he argued that the decision on whether or not the project would collapse was not in his or Morton's hands but those of the contractors. In an interview with the *Financial Times*, he said: "There is no sound reason for the project to collapse. I do not think it sensible to put at risk a project of this magnitude for what appears to be matters of personality."

Morton, for his part, made it clear that he thought the personal attacks on him were merely a smokescreen by the contractor to hide the real problem – that they were being held to a contract that they drew up and they did not like it because it was starting to cost money.

When the French courts ruled that Eurotunnel should pay the contractor £60 million as its monthly payment, the crisis was starting to reach its peak. Eurotunnel did not have £60 million. It was on the verge of insolvency.

By now Eurotunnel was signing one invoice at a time, carefully checking whether it had enough money to honour its cheques. It was down to its last £1 million – and for a company that size it was therefore pretty nearly bankrupt.

Robin Leigh-Pemberton, Governor of the Bank of England, then stepped in. In a three-hour meeting he used his skill firstly to knock sense into both sides and then cajole and encourage them to sort out their differences.

Leigh-Pemberton had good reason to get involved. The Bank of England was determined not to let the project die, having been vital in the inception of the whole affair. The Bank had backed the project when other banks and financial institutions had wavered. Its reputation was on the line and it was not going to let petty personality politics embarrass the Old Lady of Threadneedle Street.

Leigh-Pemberton proved to be the catalyst for progress in the dispute. His authority and diplomacy were well employed in helping both TML and Eurotunnel to see some sense and save the project. There was a feeling that what had begun as a traditional show of brinkmanship and bravado had got out of hand. Instead of taking the project to the edge, the two protagonists had peered a little too far over and been unable to retreat. Leigh-Pemberton turned up at the cliff top just in time.

Even with his help the negotiations were still tense. There seemed to be no easy solutions and despite confident projections that a deal would be announced on the morning of Monday February 22 1990, three days after Leigh-Pemberton's intervention, no news came.

In Eurotunnel's office above Victoria station in central London both sides continued to take this stage of the Channel Tunnel saga the full distance. TML knew how much trouble it had Eurotunnel in and was desperately keen not to miss the opportunity to take advantage. Having ostensibly started the crisis by not signing the January agreement because it was unhappy about the management situation at Eurotunnel, TML could not help wondering whether it could actually force other cost concessions out of its client. This gave weight to Morton's argument that the row was less about personalities and more about TML seeing the huge cost problems ahead and wanting to change the rules.

Eurotunnel wanted to make no more promises about fixed equipment costs, future payments or management reshuffles. Instead, it wanted to stick to the original dispute and satisfy the contractor that changes would be made in its management, as agreed as a precondition of the January deal.

Even when the five British contractors finally agreed to a deal on the Monday, the French dug their heels in once more, unhappy with aspects of the agreement drawn up in London. André Bénard worked hard in Paris on the Tuesday to allay the French contractors' fears and finally an agreement was reached, signed for TML by Neville Simms of Tarmac at 8.30 p.m. on Tuesday February 23. Simms would sign another vital agreement on the tunnel four years later, one that would again save the project. But the key to this 1990 deal was the appointment of a Eurotunnel project chief executive.

John Neerhout, a 50-year-old executive vice president of American construction giant Bechtel, was the man for the job. His most important function, as far as the contractor was concerned, was that he would act as a buffer between the contractor and Morton.

His appointment meant the loss of Tony Ridley – the project managing director – who stepped down reluctantly but received a payoff of around £200,000. Morton survived – and kept the role of chief executive. The banks relented and released the money. TML was paid. The tunnelling kept going and the project survived. But only just.

12 Tragedy strikes

"Down there the rule is get the job done. It's push, push, push, all the time."

A tunneller

Nineteen-year-old Andrew McKenna worked as a surveyor's assistant on the Channel Tunnel. He was one of the thousands of workers who every day clocked on at the Folkestone site as part of the team helping to inch the project closer to France. His job was not especially skilled. In the quaint language of the male-dominated construction industry he was more likely to be known as a "chain boy" than a surveyor's assistant. And apart from his boss, not many of the workers underground would have even known his name.

He would have travelled with his surveyor to his working area each day, clutching his retractable measuring staff and, no doubt, all the surveyor's equipment as well. When he got to his work site he would hold the staff on chosen points to enable the surveyor to measure ground levels to ensure that excavation and backfilling were being carried out to the proper coordinates.

It was mundane work for young McKenna. But on January 23 1989 an unremarkable man doing an unremarkable job became the most important statistic on the project to date. Andrew McKenna was the first man to die on the Channel Tunnel.

It was a Monday morning and McKenna was due to catch the workers' train to his work site as usual. But something went wrong. McKenna missed the train. Maybe he was late; maybe he went back to pick up more equipment. Either way, he chose to walk along the service tunnel to work. He walked on one of the railway tracks laid to supply and remove material and people in the tunnel.

The track would have been wet and muddy and very slippery. McKenna was 900 metres into the tunnel when a train approached him pulling ten wagonloads of spoil away from the working face. The train was travelling at around 10 mph and sporting a flashing light. The driver – shocked to see a figure coming towards him just 10 metres away in the murky tunnel – hit his brakes and klaxon at the same time.

A combination of unfortunate factors conspired to cause McKenna's death. As the spoil train approached, another train travelling in the opposite direction – from the tunnel portal – arrived at the spot at the same time. McKenna would have had almost no room for manoeuvre between the trains. And he would have found it extremely difficult to keep his footing on the slippery bed of the tunnel.

The spoil train hit him but was still unable to come to a halt for 60 metres, inflicting injuries on the teenager that were so horrific that the driver and his electrician were treated for shock.

There would be nine more fatalities on the project before the opening in 1994. Nine more horrifying tales of good working men who lost their lives digging the tunnel. Add to these the countless other accidents on the scheme – the Health and Safety Executive said there were 327 reportable accidents between March 1987 and December 1989 on the UK side alone – and it becomes clear that the most prestigious civil engineering project in the world took its toll on the workforce.

The health and safety aspect of the Channel Tunnel in some ways mirrors every other facet of the project. It is – like the construction, the finance and the politics – a tale of a project plunged into deep despair but finally emerging with some credit.

At one point the tunnel was losing a man every two months in tragic circumstances. At this time workers were accusing TML of "push, push, push" to get the project finished at the expense of safety.

Seven times Translink, the UK arm of TML, has been fined for safety breaches on the scheme – fines that totalled more than half a million pounds. But a rethink and a review turned the safety record of the project around until – with one major exception – its record for the final three and a half years was exemplary.

In the gruesome and sometime macabre language of construction, a tunnel job is "expected" to kill one man per mile of tunnel.

This figure was loosely based on the health and safety record of Japan's Seikan tunnel, where a 33-mile bore was constructed killing 34 men. The construction industry has a fatalistic way of accepting such statistics as "part of the job", and until recently the mentality that "construction is dangerous – if you don't like it get out" was prevalent the world over. But even in the six years since the Channel Tunnel construction began there has been a shift in attitude towards accidents on site.

American construction giant Bechtel – a company which lent top construction gurus like Eurotunnel project chief executive John Neerhout to the Channel Tunnel – has introduced a "zero accident" strategy on its sites today. Chief executive Riley P. Bechtel has set up a work ethos that gears everything to safety and refuses to accept that such a thing as an "accident" actually exists. Every incident is avoidable. His success rate, he says, has been remarkable, with 83 per cent of his sites recording no accidents in 1993. And in the United States, where the accident rate is appalling, that is an excellent performance.

Bechtel's programme demands that health and safety are put first when any decision is made on a construction site. This is backed, from the top downwards, by the company's management, which has been willing to support the move with hard cash. Bechtel recognises that good health and safety is good business.

But in the UK we have struggled to move forward consistently in the battle to reduce deaths on site. About 100 construction workers a year are killed and thousands more injured. But there are signs of change, and new regulations being introduced in 1994 will force every person involved in a construction project from inception to completion – and indeed eventual demolition – to think safety first.

When the excavation of the Channel Tunnel began in 1987 there was no doubt that the contractor was well aware of the safety implications of building a tunnel. But knowledge of the problem, and even the implementation of systems to tackle the problem, did not prove enough in the early years of work. The difficulty was in spreading the awareness of the dangers to every one of the thousands of people who worked on the tunnel. Many people close to the scheme praised the safety systems that were in place, but ensuring that these systems were employed by the many tunnel workers became the challenge. And, of course, while having systems is useful, if the work ethic on site does not stress safety then the systems are pointless.

At the Channel Tunnel, as 1989 began, there had been no deaths at all. But there had been accidents. TML knew that at the Seikan tunnel most of the accidents were not caused by tunnel collapses or floods or fire, but involved the delivery trains. And in 1988 two warning shots had been fired across TML's bow when train accidents in the Channel Tunnel resulted in two prosecutions and two fines totalling £5750 for each UK member company. The first fine – in July 1988 – was over an incident in which four empty wagons broke away and ran 300 metres to the bottom of an access tunnel. The second was over an out-of-control locomotive which hit and punctured a cylinder of liquefied petroleum gas. Fortunately, no-one was injured in either accident. But just a week before the first death on site, 21-year- old Brendan Hurley was seriously hurt when he was pinned to the track by a train.

Despite these warnings from the site and the courts, and the experience of the Seikan tunnel, it was almost inevitable that it would be the trains that were involved in the first tunnel death. The accident was a gruesome start to 1989 for the workers on site, for Andrew McKenna's relatives and friends, for Transmanche Link and for the project.

At this time TML had spent a year struggling with its tunnelling and was desperately keen to pick up pace on site. When the fatality occurred the contractor had just posted some good news on the tunnelling for the first time in several months: tunnelling rates had topped 200m a week on the service tunnel breaking all previous records and raising hopes that time might be clawed back. But this news, combined with McKenna's death, immediately gave rise to accusations that safety standards were being cut to boost speed underground.

It was an obvious conclusion for the public, the politicians and the unions to jump to, but one not backed at this time by the Health and Safety Executive, which investigated the death. In truth it would have been impossible to boost tunnelling speed to the record levels achieved in January 1989 by removing safety precautions. The improvement was due to a combination of factors, not least a fundamental change in ground conditions. But it was relevant that the more tunnelling that was carried out per week, the more workers there would be in the tunnel and more trains delivering and removing in a confined space. And the efficiency of this train system was a vital factor in tunnel progress. However fast the tunnelling machines might move, they would be hampered by any delays in removing spoil from the workface and delivering tunnel lining segments to support the excavation.

And this was down to the locomotive system.

As the tunnelling machines increased pace, so the trains had to start moving faster, more frequently and to a crucial but tight timetable. And these trains were the biggest risk to worker safety. The warning signs were there for TML and there was clearly a lot to learn on site as the pace picked up.

Worse was to come for TML and its workforce when, just two weeks after the first death on site, 35-year-old fitter David Simes became the second fatality when he was crushed between the tunnel boring machine and a crane on February 6 1989.

Simes was one of a team of four fitters plus a chargehand working on the midnight to 8 a.m. shift on fixing a conveyor belt. Their work was stopped at 4.30 a.m., but for some unknown reason Simes returned to the conveyor at 5.40 on his own to do more work. The man driving the segment erector crane did not know he was there and could not see him. From his position Simes could not see the crane either until it was too late.

The Health and Safety Executive's investigation into this accident led to the British arm of TML being prosecuted for breaching safety rules and fined £50,000. The firm pleaded guilty because it accepted that there was a failure by the supervisory staff to implement existing procedures before instructing men to work on the conveyor belt. This was the classic problem in the tunnel. The procedures existed but there were failures either to execute them or ensure they were executed.

The spate of accidents in early 1989 culminated in 500 workers walking off site for four days at the Isle of Grain segment manufacture plant after a colleague who suffered a torn bowel had had to wait 22 minutes for an ambulance. All of a sudden the safety aspects of the scheme were to the fore and the accidents prompted meetings between the contractor and the Health and Safety Executive's chief inspector of factories, Tony Linehan.

Union officials from construction union UCATT and from the Transport and General Workers Union condemned the increase in accidents as pressure mounted on the contractor to put its house in order.

The Labour health and safety spokesman at the time, Dr Gavin Strang, called for a team of health and safety inspectors to be based permanently on "this dangerous complex". Unfortunately, with the country looking at the tunnel for some reaction to claims of carelessness and dangerous operations, TML remained tight-lipped about the affair.

Eurotunnel compounded the problem by crassly stating that "site safety is a matter for the contractor", a client attitude now being challenged by new regulations on health and safety in the construction industry.

The contractor was criticised for its handling of the release of information over the deaths and was further knocked back when an ex-employee of TML was reported in the national press condemning the contractor's safety procedures. He claimed that workers' refuges – built to allow the site team to dodge trains in the tunnel – were full of equipment and therefore unusable. He also criticised the management of the site for being chaotic and over-staffed.

TML was forced to refute these statements publicly and reassure the workers that they were dedicated to health and safety. But whatever the contractor did to defuse the situation, the damage had been done. The country believed that the Channel Tunnel was a dangerous place to work – but, more seriously, they believed that the builders of the tunnel would stop at nothing to complete the scheme.

Too often the construction industry has been accepted as one that will have more than its fair share of deaths. Site workers, managers, directors and even those outside the industry looking in seem all too ready to believe that 100 deaths a year is an acceptable norm.

One of the problems is the persistence of macho attitudes about wearing safety equipment on site which force young people entering the industry to accept dangerous practices as normal because otherwise they look "soft". In 1989 at the Channel Tunnel the fact was that the contractor had extensive safety systems in place that would put most construction sites in the UK to shame.

But it was not enough. On a scheme as big as this and with the sheer quantity of people working within a confined space, the safety precautions had to be more than good. They had to be exceptional.

With two men dead in two weeks at the beginning of 1989, TML was starting to get an idea of how difficult it would be to protect its workforce.

The Health and Safety investigation into the accidents resulted in Translink's third prosecution, eventually heard at Maidstone Crown Court in March 1990, where the firm pleaded guilty to breaching the Health and Safety at Work Act. By the time the prosecution got to court, three more men had died in the tunnel – two in the UK and one in France. So when fines of £10,000 per member contractor of Translink were imposed, there was an outcry.

At the time it was calculated the latest profit figures for the five firms totalled more than £750 million. The then senior area director of the Health and Safety Executive Jeff Hinksman, a well respected and refreshingly candid inspector, said: "You have to ask yourselves what sort of effect a £50,000 fine has on firms of this size." It was little more than a slap on the wrist, he said.

Hinksman's annoyance was backed up by union leaders. Transport and General Workers Union construction spokesman George Henderson called for jail sentences for individuals to ram home the safety rules and condemned the fines as "totally inadequate". George Brumwell of the construction union UCATT put the blame on the management: "I've been down the tunnel and the problems are the same as those found on any site. It isn't the high tech equipment that causes the safety troubles, it's general housekeeping. If they can't get the basic housekeeping right then that highlights the poor quality of the management."

Brumwell's point about the site was right. Anyone who has visited a large construction site will know that you have to have eyes everywhere to watch for dangerous plant trundling past, while at the same time taking care not to fall into excavations or shafts or be hit by materials being handled around the site. On a dry day in the summer an outdoor site can be dangerous. On a wet day with murky light it can be lethal. Put the same plant in a noisy tunnel with little room for manoeuvre, poor light, wet conditions and no easy escape, and some very special safety precautions are required if that site is not going to become a massive death trap. And that, as Brumwell said, is where the management is so essential. Having admitted safety failings exposed by the HSE investigation following Simes' death, the contractor was fortunate to escape with such a paltry fine. But the case only concerned safety breaches made 12 months earlier. And since that time the safety record on site had deteriorated badly.

In August 1989 the safety spotlight switched to France, where 29-year-old Guy Jolly was crushed to death by segment handling machinery.

The French had generally maintained a better safety record than the UK side, a fact disputed by the British team, who argued that the reason for their lower accident rate was that the French had fewer people working on the tunnel. The atmosphere on the French side was, however, strangely relaxed. Smoking and drinking – strictly banned in the British tunnels – were allowed in France without noticeable mishaps. Union leaders say the more organised union structure in France helped keep health and safety to the fore, and in the end only two men died on the French side compared with eight in the UK. The second death was 56-year-old René St Georges, struck by a segment train in May 1990. At least, these are the official figures.

In a remarkable outburst in October 1990 TML's group safety manager Alan Sargent claimed that there had been at least three more deaths in France than stated, and that it was the vagaries of the French reporting system that kept their death toll to two.

Sargent claimed that two workers had been killed in August 1986, swept overboard from a vessel while carrying out exploratory work, and two men had died in road accidents at the site. Another had been overcome by fumes in his site caravan.

He told an Institution of Occupational Safety and Health meeting: "I get fed up with these (British and French) comparisons. We are building two thirds of the tunnel to their third and we employ more men." But he accepted that the attitude to safety of the average tunneller was minimal.

The remarkable French death claims were not well received by TML's management, and within a month Sargent had left the project, saying he had been made a "scapegoat". It was a move that upset the workforce, who respected him and believed he listened to them.

The third UK fatality came in October 1989, just two months after the death in France of Guy Jolly. Gary Woodward, a 32-year-old father of two from Sheffield, was crushed to death between the side of the tunnel boring machine and a hydraulic concrete segment handling tray.

Woodward was fixing a piece of track into position in a main undersea running tunnel when the tray – which lifted segments into position – sparked into life involuntarily and began its automatic cycle of operation, pinning the man against the machine. The inquest found that he died of crushing and asphyxia.

The only possible explanation of the accident was that a power line was nipped by a guard rail on a nearby segment erector crane, sending the power surge to start the operation.

Again the contractor was prosecuted for failing to ensure the safety of its employees, and this time joined in the dock by boring machine manufacturers Robbins and Markham who were charged with "failing to provide a boring machine in such a way as to ensure workers were not exposed to risks to their safety." All seven companies, the five members of Translink and the two tunnel firms, pleaded guilty and were fined £6000 each in September 1991.

It is worth noting that although Woodward's death was one of seven on site in an 18-month period, there was little complaint about the level of fine in his case because it was not heard until September 1991 – when safety was much less of a hot topic on the tunnel.

The Simes case, on the other hand, was heard at a time when accidents were all too common on the tunnel, so was more of an issue and attracted industry and public outrage.

Despite these complaints the fines given out remained low and did not seem to take into account the whole picture. Not only were there two fires in the tunnel around the time of Woodward's death – caused by problems in the spoil locomotives and resulting in minor injuries to nine workers – but the incident was the start of the tunnel's worst safety period. Deaths would follow in January and March 1990, with two more in May – one in the UK and the loss of René St Georges in France. These deaths sparked a crisis on site and forced three safety audits of the project to be carried out. While the crisis meetings continued, the scheme claimed its ninth victim in July.

It began on January 12, when 34-year-old grouter Keith Lynch from Retford in Nottinghamshire was struck by a six-car rail track delivery train in the same tunnel where Woodward had died. The trains were driven from the locomotive at the back, pushing the wagons ahead rather than pulling them. On that day learner driver Paul Davies – on his second day operating a loco – was in charge of the train.

The inquest into the case heard that although there were banksmen employed to wave traffic on in the tunnel when it was safe, there had previously also been a worker employed to sit on the front of these trains and communicate with the driver by radio. This banksman on the loco, the inquest was told, was not present that day.

As Davies pulled off with his load he had only been travelling a few yards when he saw people yelling for him to stop. His train had struck and killed Lynch, who had been bending down to recover a grout plug from the tunnel lining and was not visible to the driver, despite wearing a fluorescent yellow jacket. The incident sparked a furious letter from George Henderson to the HSE which said: "I am at the end of my tether, feeling angry, somewhat frustrated, deeply shocked and numbed. What the hell is going on?" Translink was eventually fined £125,000 over the accident.

Hardly had the ink on the letter dried when Stephen Wright, a 40-year-old Folkestone plumber, was crushed to death in April under a 300 kg pipe which he was fixing to the sea wall at the base of Shakespeare Cliff.

In this instance two men, Wright and crane operator Brian Magee, were carrying out an operation that would normally have been undertaken by three men – the third had not arrived for the day.

Magee operated the crane to lower the pipe to Wright, who was working on the sea wall, fixing the massive steel drainage pipe into place in brackets. It was standard practice to leave the sections of pipe loosely attached so that the next piece could be fitted more easily. The pipe section fell and crushed Wright beneath it. He died of internal haemorrhaging and compression of the chest.

The two deaths had caused anger in many quarters, but when 33-year-old grouter William Cartman was hit and killed by a segment erector crane a month later in May, there was mayhem. Cartman was working near the cutting face of the boring machine, ensuring that the grout was setting. The segment erector operator did not see him and he was struck by the crane. The coroner said his death was caused by "carelessness, neglect and bad luck", and in April 1992 the contractor was fined £90,000 after admitting safety breaches regarding the accident.

Work was halted on the site after Cartman's death and Employment Secretary Michael Howard called TML chief executive Jack Lemley in to see him to explain the accident rate on site. It was a rare interference in the project by the government, but public concern was such that politicians of all hues were calling for full inquiries into the problems. Indeed, the contractor had to report to a Transport Select Committee on the health and safety crisis.

The Health and Safety Executive began a major audit of the project's safety procedures, which TML insisted were in order. The problem, which TML recognised, was getting the workers to abide by the rules. Union safety representatives were operating on site but their numbers were small compared with the thousands of workers in the tunnels.

TML continued to maintain that the increased tunnelling speed was not the cause of the sudden spate of accidents. There was no pattern to the accidents, said the contractor. They were generally caused by carelessness and complacency under the ground. But confirming TML's arguments with the site workers was particularly hard. It was difficult to find someone at the sharp end to describe the dangers of the work in the tunnels. But in an anonymous interview with the *Independent on Sunday*, tunnellers said what they thought about conditions, angered by the deaths on site. "It is the worst tunnel job I've worked on. It's a massive project and there are lots of safety rules and regulations, but they don't operate when you're actually down there," said one. Another condemned the facade of the induction training course. "Before you start the job there is an intensive four-day induction course," he said. "But the minute you get in on your first day in the hole you can forget the induction. They tell you never to walk on the tracks, then as soon as you step out at the pit bottom the first thing you come to is a set of tracks. You have to walk on them. Down there the rule is get the job done. It's push, push, push, all the time."

The men said they were not allowed to complain. If they did they were branded as troublemakers and given worse, less well paid jobs.

These were damning criticisms, which TML vociferously denied. But a site clampdown on talking publicly and the secrecy over the health and safety audits all lent weight to the idea that something underground was amiss.

The conclusions of the Health and Safety report were, however, made public. The report – while being supportive of TML's safety efforts on site – said there was the risk of "catastrophic loss" and listed the "more significant matters for concern." These included the fact that more than half the management had not attended the induction course and that the managers were not held individually responsible for health and safety. Management was also criticised for not attending safety meetings or showing positive attitudes towards safety and health. The HSE also picked up on the lack of workers' cooperation and that there was no system for workers to show their concern about safety. The report said "the relatively low result in leadership and management is a notable weakness."

TML chief executive Jack Lemley welcomed the report and promised to act on its specific recommendations as soon as possible. It was an important turning point for the contractor, who used the opportunity to change attitudes on site and improve the situation in the tunnel and with its managers at all levels.

TML also found itself facing angry union chiefs in a meeting to try to clear the air about the alleged victimisation of workers who complained about conditions in the tunnel.

The meeting had been a long-standing arrangement between the TGWU and TML and set for August 1 1990 to discuss the HSE recommendations and the safety problems on site. Three days before the meeting, the ninth worker on site died. Charles McCourt, a 45-year-old grandfather from Oldham, who was married with three children, was electrocuted when carrying out repairs to a tunnel boring machine. This death brought Paul Gallagher of the electricians' union EETPU on site to join the TGWU meeting with TML. Gallagher brought up the issue of intimidation of workers, and was backed up by Colin Campbell of the TGWU. Campbell said: "For a worker to say: 'That's against the procedures and I can't do that,' and the foreman to say: 'Yes you will or I'll find somebody else,' and the worker then to be forced to breach procedures – that's what we call intimidation."

In a heated meeting TML maintained that such intimidation was not taking place and that specific cases had been investigated and no substance found. But the union chiefs persisted with their complaints and TML promised to look again at the situation.

Paul Gallagher said: "I am sure that senior management are committed to safety. But I am not sure that commitment goes down the management ladder."

The unions, the HSE, and in some ways the contractor, were in agreement that the problem was getting the safety message down the line to everyone – but particularly agents and foremen who told the workers in the tunnel what to do.

Part of the problem was to identify whether the section bosses were pushing the workers too hard, and if so why? Were they being put under too much pressure themselves by their bosses to finish work quickly?

The deaths had caused everyone to concentrate their minds on the problem and, with a change in attitude on site, it was seen as possible that accidents due to carelessness or complacency could be stopped.

TML scrapped its safety management system and, on the advice of the HSE and under the American influence of Jack Lemley, introduced an American-style safety system to the site.

The system is popular in the US petrochemical industry and was developed by the massive Dupont company, which prides itself on its safety. Indeed, specialist Dupont Safety Management Services was brought on site to boost safety.

The system is all about gearing everything and everybody on site to safety and to engender a culture change in attitudes, from the top of the management to the bottom, so that safety comes first, second and third. It is similar to the zero-accident policy introduced by Bechtel and was warmly received in all quarters.

John King, tunnelling director at TML from 1988 to 1990, said that TML realised that things had to change on site. "The Dupont system changed everything. I can't stress strongly enough how it introduced a culture change which meant that the first thing you ever spoke about on site was safety. It required a lot of training sessions for the supervisors from top to bottom. Eventually safety was in everybody's mind all the time."

The introduction of the system was part of a vast remodelling of TML's attitude to safety. The project began to be managed for safety like the best of the big American sites. Videos and posters backed up "tool box" talks before shifts, and the number of safety reps was boosted and each encouraged to be active in their policing of the work.

The system changed attitudes on site and, while not able to eliminate accidents, reduced them dramatically and stopped the depressing sequence of deaths at a stroke. The unions and management started pulling the same way, and for more than two years ran the project without a fatality – a major step forward after the deaths of nine men in 18 months.

It was all the more surprising, then, when in October 1992 David Griffiths, 26, was killed while working as a banksman on a train in the tunnel.

The accident was described by Judge Felix Waley at Maidstone Crown Court as the "worst failing" the court had heard over the last few years. He said: "It is inconceivable that the companies were unaware of the fact that the safety procedures in relation to the narrow-gauge railway operations were not being followed properly whatever the difficulties. This accident happened because safety procedures were not properly in place and properly supervised."

Griffiths, who had had no training at all, had been appointed banksman the week before the accident and was operating the points when he was crushed between a hygiene train and a stationary train alongside. Griffiths had only one eye, and when he was first employed in 1990 he was declared only able to do surface work, but this restriction was relaxed in February 1992.

The five companies were each fined £40,000 and ordered to pay £3000 costs. This was the biggest fine imposed on the contractor and a sorry end to a project where the worst and the best of construction health and safety were exposed.

For the ten families devastated by the project, nothing will convince them that the tunnel was a good thing. But the positive strides made by the contractor towards setting up the world's top safety systems on a British site will go a long way to reducing construction deaths in the future.

13 The £1 billion row

" Unless there is a considerable change in attitude by Eurotunnel, the scheme will come in late."

A TML source

When John Neerhout took the job of project chief executive at Eurotunnel in February 1990, the row over costs had been running for well over a year. The estimated final bill for the scheme had risen to around £7.2 billion – although contractor and client still could not agree on this – from £4.9 billion. The contract was a legal mess which would inevitably lead to more rows, and four men had died on the scheme in a year. The contractor had been forced to make management changes, as had Eurotunnel, a matter both sides were still smarting over. TML and Eurotunnel were also both convinced that the other side had acted appallingly during the crisis and had brought the world's best known civil engineering project into disrepute. The only good news was that tunnelling was improving, but even that was still behind schedule.

The Channel Tunnel project was in serious disarray. To make matters worse, British and French pride had been dented. With Jack Lemley firmly in place as chief executive of TML, and now Neerhout taking over the project reins at Eurotunnel, the Americans seemed to have hijacked a great European venture. From the outside it certainly looked like the cavalry had been called in to clear up the mess left by the British and French. There was some criticism that construction practices on this side of the Atlantic had led to the problems and that the enlightened American view was the way forward. But in fact the project's squabbles knew no national boundaries and it became apparent that Neerhout and Lemley did not get on with each other either.

However, in 1990 the appointment of Americans to manage the project was seen by many as an indictment on the British and French abilities to manage a major construction contract.

Neerhout was not a stranger to the scheme when he was appointed.

Many of his Bechtel colleagues were seconded to Eurotunnel and he had been across several times to talk to Morton and Bénard about how to handle the contractor. So it did not take him long to get his feet under the desk and begin the task of guiding the project through to a successful completion. He certainly did not take long to make an assessment of what was wrong with the scheme, and in an interview with *Construction News* he put his finger on what he saw as the key problem on the project – the lack of clear design. "My priority task is to finalise the scope of the project. Some things, such as the signalling and the smoke protection systems, have not been finalised yet. Once these items are resolved then the design is frozen. Then the estimators can tell you what it is going to cost. And once you know what it is going to cost, then you can organise your financing. So I want to get the design scope of the project finalised quickly...I am convinced we can get it built, but we have got to get it designed first."

This design uncertainty, Neerhout knew, was the nub of the problems between contractor and client. What it affected more than anything was the fixed equipment – the power, ventilation, cooling and signalling systems that had to be fitted into the completed tunnel to allow a full-scale extremely busy railway to run smoothly under the English Channel. The problem was that the details had never been pinpointed. Without this information the 10 contractors had put a price on the installation of the fixed equipment which never had a chance of being close to the correct sum.

This had been done before Eurotunnel existed. It would have been impossible for the tunnel promoters to add in the possible costs of future delays and difficulties. So the figures that the banks saw represented the cost if work went relatively well. Once the banks began to lend the money they were hooked, and it was difficult to pull out as costs increased.

TML vehemently denied that it had underestimated the cost of the work in the beginning, arguing that independent institutions had agreed with its cost forecasts. But the amount of detail on the specification of the system was so scanty in those pre-construction days that calculating its cost was an impossible task – the designers might as well have used the back of an envelope. What is surprising is that the banks missed this point when they began funding the project.

So by early 1990, with costs escalating but with no agreement by how much, the banks began asking what the difficulty was. They were told about the lack of design, and quite reasonably argued that surely the design was agreed in the original concession. Neerhout, coming into the cost crisis from afar, had a neat analogy to explain this discrepancy: "The banks say the design should be in the concession agreement. And you say 'sure it's in the concession. It says "it's a kitchen" The question is what sort of kitchen?' Your wife might have figured on marble counter tops and the contractors on formica....that's what we have to sort out."

Neerhout also recognised that sorting this out would not be easy because the financial discrepancy between the two sides was so huge. Already there had been an agreement to slow the trains through the tunnel to downgrade the systems required to operate the scheme, but that was not enough. Other ways of cutting costs were still being explored. Anything ornate and aesthetic but not functional went out of the window as both sides struggled to bring costs down.

It was becoming increasingly clear that the cost difference between the two sides was not only immense, but growing by the day. What was refreshing was this did not affect Neerhout. He argued that once the design was settled a cost could be settled. If there was a continuing dispute over the costs, which looked certain, then the contract allowed for a disputes procedure which would rule on the claims, and a right to go to arbitration if necessary. What concerned Neerhout was the management structure of the project, which struck him as over-complex. He told the *Sunday Times* that the work that had to be done was relatively straightforward. It was the complexity of the operating structure that worried him. "It is very difficult doing a job with two contractors joint venturing. It's nearly impossible with three unless one is appointed the lead contractor. But when you have got ten together with equal authority in two different languages, it's very difficult to sort out."

What Neerhout wanted to ensure was that his opposite number at TML, Jack Lemley, was given the power to act on behalf of the consortium. During the tense negotiations over the January deal Eurotunnel asked that Lemley be given that power so he could deal with Neerhout. This meant that the big guns at the individual contractors – such as Martin Bouygues of Bouygues, Cliff Chetwood at Wimpey and Peter Costain – would take more of a back-seat role while the two Americans made some big decisions.

There were no easy solutions to the crisis. There is no miraculous story of the two men sitting down that summer and cracking a quick deal to ease the project safely to completion. But what they did achieve for a while was to calm down the scheme and remove the costs dispute from the intense media spotlight. This allowed the two sides to have a genuine shot at sorting out the difficulties and finalising the design without a war of words being conducted through the media. This formula was sufficiently successful for Eurotunnel to come through the sensitive announcement of its 1989 financial results in April 1990 relatively unscathed.

At this announcement Morton revealed that Eurotunnel was talking to its funding banks about raising a further £2 billion in loans, with another £500 million likely to be raised through a rights issue later in the year. This would put Eurotunnel's available funds at £8.5 billion – about £1 billion higher than their estimates of the cost of the project.

That such a statement could be made without too much fuss was a reflection on the inability of the project to shock people any more. Unless the project was on the verge of collapse – again – people were willing to believe that in Channel Tunnel terms everything was progressing relatively smoothly.

On the costs, Morton explained that the fixed equipment claim was not yet settled. This was of little surprise, given Neerhout's assessment of the situation just two months earlier. "From the scanty information made available to us by TML regarding their intentions," said Morton, "it seems they have in mind sums aggregating in excess of £700 million." This was considerably higher than the £380 million gap between the two sides revealed six months previously. But there was no explanation why that gap was increasing.

Morton expressed more concern over tunnelling, which he said was currently expected to cost £500 million more than expected. This, he said, was due mainly to the increase in the number of workers TML needed to complete the work.

Whether Eurotunnel consciously played down the fixed equipment dispute is not clear. But it did tend to put it to one side as the company began to work to convince the 220 funding banks to bail the project out.

Eurotunnel was pretty much resigned to seeing the £700 million fixed equipment claims dispute dealt with under the contract's existing disputes procedure. It was quietly confident that TML would not be able to back up its claims in the detail required to justify the multi-million-pound cost increases. Eurotunnel also knew that even if TML had a good case, the practical difficulties of putting together one of the biggest claims the construction industry had ever seen would be insurmountable.

Although site negotiations about the costs continued every day at varying levels of management, Eurotunnel's big push was for new funding. Without that, after all, there would be no project to argue about.

Initial soundings were positive. The banks' reaction to the news that Eurotunnel was looking for another £2.5 billion in loans and equity seemed to be that they would help this one last time. The City of London analysts felt that the key was convincing the banks that the project was sufficiently under control that Eurotunnel would not be running back in 1992 asking for yet more money. But Morton had said quite clearly at the results press conference that this would be the final fund-raising exercise. "We don't expect to go through this again," he said. But as the summer of lobbying and cajoling banks all around the world wore on, doubts began to emerge that the banks would indeed back the project quite so happily. And in August Eurotunnel dropped the bombshell that once again plunged the Channel Tunnel into crisis.

Fewer than half the 220 funding banks had agreed to lend more money to the project. As a result little more than half of the £2 billion of the additional lending it was seeking had been secured. The refinancing package required Eurotunnel to have the extra £2 billion in place by the end of August to allow it to proceed with a £500 million rights issue in October. As of mid-August only 91 banks had said "yes" to a further loan, 93 had said "no" and the rest had not replied.

Morton said that the climate for bankers was not good and explained that the banks were balking at the amount of money required. With costs rising dramatically on the project and TML still lining up a £700 million claim, it was becoming increasingly difficult to convince the banks that further backing was a good idea.

Eurotunnel blamed TML's 1989 tunnelling progress and its pursuit of its £700 million claim for scaring the banks. This of course was true, if disingenuous. If TML felt it was owed that money, it had every right to ask for it whatever the financial situation of its client.

The banks' concern was that this would not be the last request for extra funding and that Eurotunnel would sooner or later be back to ask again. And despite all Eurotunnel's protestations to the contrary, they were right in the end. Eurotunnel would eventually require yet more money to keep the job going.

There were other factors that summer that conspired to hinder Eurotunnel's progress in securing the loans. Iraq's invasion of Kuwait had unnerved many world banks, but particularly those in the Middle East. In Japan, the Tokyo stock exchange had fallen sharply and interest rates were high, discouraging many of the banks from helping.

The refinancing hold-up was starting to cause cash flow problems on the project again. Although Eurotunnel had by no means spent all of the £5 billion of loans already secured for the contract, the money was not readily available for spending. Instead, Eurotunnel had regularly to "draw down" part of that sum, and the banks only released portions of this money on certain conditions. One of those conditions was that the total funds available to Eurotunnel must be sufficient to cover the estimated cost to completion of the scheme. That cost to completion figure had risen substantially, forcing Eurotunnel to find more money. But without the £2 billion loan, Eurotunnel was in breach of its agreement with the banks because its available funding did not cover the estimated cost to completion.

At any time the banks could refuse "draw-downs" on the original loan, so cutting the cash flow to the project and meaning that Eurotunnel would have to stop payments to TML on site. Eurotunnel was therefore requesting a waiver of these rules to allow funds to continue. "A refusal from the banks is a vote to stop the contract," said Morton.

After negotiation a waiver was granted while Eurotunnel chiefs dashed around the world convincing and cajoling banks to change their minds about the loans. A £300 million loan from the European Investment Bank boosted the project, but still the money fell £600 million short of that required.

(Above) A tunnel miner is given a helpful hint about which way the project should go in 1990.
(Left) The British terminal at Cheriton under construction.
(Below) The project would not have been built without the backing of Margaret Thatcher, here seen with Eurotunnel co-chairmen André Bénard (left) and Alastair Morton.

(Above) Tunnellers service the running tunnel machine as it passes through the crossover chamber.
(Right) Tunnellers work below the ghoulish spectre of the massive north running tunnel machine.

(Top left) 30th October 1990 and the probe sent out from the UK service tunnel boring machine makes contact with the French tunnel miners.
(Top right) The moment of breakthrough in December 1990. French miner Phillippe Cozette greets Briton Graham Fagg.
(Bottom right) TML's team celebrates the first breakthrough under the Channel in December 1990. From left to right, Construction Director Peter Allwood, Tunnels Director John King, Chairman Phillippe Essig, Chief Executive Jack Lemley.

(Top left) Murphy's marauders lead the way in the landward break-through of the UK's south running tunnel.
(Top right) Paddington Bear and bottles of champagne help the first breakthrough party go with a swing.
(Right) Miners celebrate the third and final under sea breakthrough, the south running tunnel, in June 1991.

The deadline to have the extra £2 billion raised passed by at the end of August, forcing the imminent rights issue to be put back, and still Eurotunnel struggled to woo the banks to back the project.

On site, TML was concerned about the situation. Only the banks' goodwill in granting the waiver was allowing payment to come through each month. TML was also worried about placing long-term construction orders without knowing that the funds were coming through.

As the crisis rolled on the progress on site continued to be remarkable, and this in many ways was the factor that convinced enough banks to dig deep to keep the tunnel alive. This was rather ironic, as according to Eurotunnel the funding crisis had been caused by poor progress on site in the first place.

As the battle to convince the banks moved into the autumn it became increasingly apparent that the service tunnel was going to break through on schedule at the beginning of December 1990, connecting Britain and France for the first time since the Ice Age. The boost that such a breakthrough would give to the project would engender a new wave of optimism on site, in the City and among the public.

Some of the banks recognised this and the lobbyists and Channel Tunnel supporters played this card for all it was worth. One by one they convinced the bankers that the project was a good bet. They described the sensational tunnelling progress and asked whether – with breakthrough imminent – this was a good time to jeopardise the tunnel's future. Eurotunnel did finally convince enough of the world's banks to back the scheme again – but only just.

Sure enough, in October 1990 Eurotunnel was thrown the lifeline of a loan of £2.1 billion, more than the £2 billion target set by the banks as a condition for going forward with a rights issue – now scheduled for November.

Technically the money had not all come from the 220-strong banking syndicate. The £300 million put up by the European Investment Bank meant the syndicate was actually £200 million short of the £2 billion target. But in the circumstances Eurotunnel had performed wonders. Many banks from the original syndicate had refused to lend more money, particularly some Third World banks. But surprisingly some French and German backers had refused to support the project further. The British had come through strongly, but the key had been convincing the Japanese, who had been subjected to some high-level pressure to keep the project going.

This financial breakthrough was followed days later by the undersea service tunnel breakthrough. The British and French service tunnelling machines, guided by laser, had bored to within 90 metres of each other and a 50 mm diameter probe was sent out from the British service tunnel machine to make contact with the French machine more than 50 metres

EVENING STANDARD, 13 JUNE, 1990

"Oh, look! They're going to do another estimate!"

below the sea, 22 km from Shakespeare Cliff and just over 15 km from Sangatte. Although it would be another month before tunnellers from France and Britain could shake hands, the moment the probe touched that French tunnel machine was technically the first land connection between Britain and France since the Ice Age.

It was a great moment for all those involved, not just from an historical viewpoint but from an engineering aspect as well. Laser-controlled the TBMs may have been, but the fear that the machines will be off line and not meet underground is always present in tunnellers' minds, as Eurotunnel's technical director Colin Kirkland told *Construction News* at the time: "Breakthrough is always a tense time even on the smallest tunnel. But when you've dug tunnels as far as these two machines have gone it's particularly scary. There is always that element of doubt, although we were quite confident the chap would be there on the French side."

The confidence was well founded and the machine and probe came into contact well within the 150 mm tolerance allowed, to the delight of all.

The breakthrough did throw up one difficulty which had not been fully considered before work had started – what to do with the tunnel boring machines once they had finished their work. In the end the decision was made – once the probe was sent out – to drive the cutting head of the

British machine off line and away from the tunnel to be entombed in concrete 50 metres under the sea. The back-up machinery could of course be removed.

The final plug of chalk between where the British machine had finished its work and the French machine stood was then excavated by mechanical digger and the French machine dismantled and removed from the excavations – an arrangement planned by the French with a touch more foresight than the British.

This technique was broadly repeated for the running tunnels, so three massive tunnel cutting heads are now entombed below the sea.

All in all it was a remarkable month for Eurotunnel. The money was in place and the tunnels were on line and on schedule. So it was time for a new row.

The summer's task of obtaining new funding for the project and the excitement of the tunnelling progress had detracted from the big problem still facing the project: costs. Neerhout and Lemley had kept their heads down and worked towards agreement on the final price of the tunnel. But when details emerged in November 1990 of the state of play it was clear there had been no progress and that the two Americans were far from a deal. In fact the situation had deteriorated.

Eurotunnel's success in rounding up the £2.1 billion loan for the project gave the client the go-ahead to launch a rights issue aimed at raising £500 million. With this in mind, TML was holding back on its claim to try to smoothe the path to a successful launch by Eurotunnel. But Eurotunnel was bound to produce a prospectus before wooing the shareholders, and this required publishing some home truths about what was owed and what was claimed on the scheme.

As was now the norm with this fascinating project the prospectus – published on November 5, six days after the historic undersea tunnel connection – certainly made interesting reading. It revealed that TML had lodged a massive number of claims with Eurotunnel which had not yet been settled. In all 114 claims were outstanding – adding up to £953 million. A further 87 claims had been agreed. Of the £953 million the vast proportion – £811 million – related to the fixed equipment, the part Neerhout recognised as a potential hazard the day he took office.

The prospectus said that the company was sceptical of the basis of many of TML's claims, as no details had yet been given of how the totals had been arrived at by TML. It said that it accepted so far that TML was entitled to an extra £209 million for work carried out on top of the original specification. It also told prospective shareholders that it had established a project contingency to cover extra payments above this level. But it was bound to add: "It is not possible to make a definitive assessment of TML's claims. Nevertheless, the directors believe that the assessments and provisions that they have made are reasonable. However, if these assessments prove to be materially wrong, or if the provisions prove insufficient, this could have a material impact on Eurotunnel's financial position."

TML accepted that many of the claims had not yet been substantiated and that the total could yet be reduced by negotiation. But the contractor knew that with the fitting out about to start in the service tunnel in the New Year, the design delays had caused confusion and even an element of chaos in the planning and procurement of the mechanical and electrical systems. The situation was going to get worse not better.

The prospectus had a second shock for Eurotunnel's shareholders – both existing and prospective. TML had submitted a 55-week extension of time claim on the project for the development of the signalling system. This did not mean that TML thought the tunnel was going to open more than a year late, but it did mean that crucial sections of the work were taking longer than expected, and to speed up the work to get the tunnel finished on time – in June 1993 – would be very expensive.

What TML wanted was the deadline to be pushed back, so that when the tunnel was completed on time it would receive bonuses for effectively finishing early. As things stood, TML was late and facing penalties for the delay.

Despite the doom and gloom in the prospectus, the excitement of the project was such that by December 3 the rights issue was completed and raised a further £566 million.

For Eurotunnel the rights issue had been brilliantly timed. Despite the clear difficulties emerging in the project, people were carried away with the euphoria of the service tunnel breakthrough. This was covered live on television, a remarkably high profile for a construction project. The industry had never been the subject of so much interest. The historic moment was caught on camera as, commendably, Eurotunnel and TML decided that the key meeting and handshake below the sea would be between tunnellers. Not politicians, personalities or top executives, but two of the guys who had actually dug the tunnel.

Tunneller Graham Fagg from Dover met French counterpart Phillippe Cozette through a small hand-dug hole, swapped flags and linked Britain and France. Both sides enjoyed an almighty party – although not surprisingly they did still manage to disagree about how it should be held and who should be there. The tunnel, after all, still belonged to TML, and the contractors wanted the party to be as much their event as Eurotunnel's. But these were petty squabbles compared with most, and for a day at least the two sides celebrated a remarkable achievement and forgot about the almost £1 billion of outstanding claims.

Not for long. It was less than a week after breakthrough that TML resurrected the subject of the cost of the fixed equipment element of the contract.

There had been a kind of truce between Eurotunnel and TML building up to the rights issue in November 1990 and the widely covered service tunnel breakthrough in December. But TML was acutely aware of the problems still developing on site despite the public perception that with the tunnelling complete the Channel Tunnel would soon be open.

Still the main worry for TML was the lack of design information coming from Eurotunnel on key areas of the mechanical and electrical work. Without this information TML was struggling to plan properly its procurement and site work, and this would eventually lead to inefficiencies on site that would cost money. This, TML and Eurotunnel both knew well, would lead to further claims and further financial wranglings on the tunnel.

While on the one hand pressing for final details of Eurotunnel's requirements, TML was also keen to start sorting out its claims to date. While Eurotunnel had spent the summer convincing banks to back the project, TML was quietly trying to piece together its massive claim for more money. But it was struggling.

In a normal lump sum contract – such as the construction of a road or an office block – variations are regularly made to the contract by the client. These are agreed by teams of quantity surveyors – or on-site cost accountants – representing both sides, client and contractor. The contractor then submits a claim for how much its quantity surveyors estimate the extra work has cost. This claim is based on the schedule of prices per unit of work which formed the basis of the contractor's original tender for the contract.

So, for instance, if the contractor had priced £10 per cubic metre of excavation in its original tender for the job, then that is the basis for a claim for extra excavation work ordered by the client. However, although it is the basis for the claim, it is unlikely that the rate will stay at £10. The contractor will argue that the original rate was for bulk excavation programmed carefully at the beginning of the job. Any call for extra work – especially in small quantities and after the rest of the excavation has been carried out – will be charged at a higher rate calculated in detail and justified by the contractor's quantity surveyor. In short, a claim for even a relatively small amount takes a long time to prepare.

TML's detailed claim for nearly £1 billion was becoming something of a nightmare. Everything had changed so fundamentally on site that it was difficult to base new rates on original rates – especially considering that these original rates were calculated in haste when the scope of the work was not understood. Going back to Neerhout's kitchen analogy, a price had been quoted for: Item: A Kitchen. Quantity: One. Price: £x. How could that price form the basis of a new claim generated by a call for extra taps, marble tops and recessed lighting? Despite this the first claims were now filtering through for discussion between Lemley and Neerhout.

At the end of 1990 Eurotunnel confirmed that the claims were not yet detailed at all, but were just figures at the bottom of a page. What the client was waiting for were the detailed costings, which would then be considered item by item. But already Eurotunnel knew that TML would struggle to produce such detailed information.

The signalling claim – as highlighted in the rights issue prospectus – was however already going forward through the contractual disputes mechanism for a ruling. This included TML's claim for a 55-week extension of time.

The claim had been submitted – as demanded by the contract – to the project disputes panel, a team of experts given power under the contract to consider any project disagreements and make rulings on who should pay, be paid or be awarded what. The signalling claim decision was due in January 1991, and with so much now at stake was awaited with some trepidation by both Eurotunnel and TML.

The importance of the claim was that if it ruled in TML's favour the contractor could in theory reprogramme work beyond the June opening date, unless it was given the detail it needed in the fixed equipment design and the leeway it needed in its financial claim.

It would also open the door for further claims costing Eurotunnel millions of pounds. If, on the other hand, the ruling was in favour of Eurotunnel, TML faced a massive bill for the fixed equipment work, with little chance of compensation.

TML told the disputes panel that in 1988 Eurotunnel had interfered with its awarding of the signalling contract for the scheme, and by doing so had delayed work on that area of the project by more than a year. TML said that it had been negotiating the contract with two contenders but Eurotunnel had stepped in and told TML to drop one of them, a consortium called Eurosignal. This, said TML, fundamentally affected negotiations for the contract, which eventually broke down, so delaying the work. TML said it had a contract to design, build and commission the Channel Tunnel – and that Eurotunnel had interfered with this contract, costing 55 weeks of time.

Somewhat naively the assumption was that one side would be ruled the winner of this dispute by the panel, the other side the loser. The press certainly wanted the situation to be black and white. In the light of this a decision by TML and Eurotunnel to impose a news blackout on the ruling was somewhat misjudged. Under an insistent media spotlight they could not hope to suppress all the information on a ruling so important to thousands of Eurotunnel shareholders.

With hindsight a joint public announcement of the ruling and a discussion of its interpretation would have been the best way forward for the project. Instead, the attempted news blackout turned into another row between Eurotunnel and TML.

The first leak of the panel ruling pointed to a humiliating defeat for the contractor. The panel had ruled against the 55-week time extension claimed by TML, and on the face of it this decision could cost TML anything up to £400 million. It also seemed to jeopardise further claims for delays on other areas of the work.

The news that TML had lost resulted in a 24p increase in Eurotunnel's shares to 499p. Unfortunately it was not the whole story. Subsequent leaks from the ruling revealed that although TML did not get its 55 weeks confirmed, the panel did agree that Eurotunnel had interfered with the signalling contract award and that it should not have done so. This put a whole new light on the panel's findings. Although it had dismissed the 55-week delay, it did not dismiss TML's right to ask for extra time because of any delay caused by Eurotunnel, and went on to rule that Eurotunnel should pay for that delay. This suddenly looked expensive for Eurotunnel if it was forced to pay for the contractor to accelerate the work to finish on the June 1993 date.

In the third leak in the contract's now farcical news blackout Eurotunnel told *The Times* that the panel ruling actually meant that there would be no increase in cost over and above that outlined in the November rights issue prospectus. This had allowed £209 million against TML's £953 million of claims, but also established a £239 million project contingency. TML was livid at this revelation which, it believed, totally misrepresented the situation.

The panel ruling had actually stirred up a hornet's nest and certainly had not solved any dispute. Asked to rule on a 55-week extension of time, it had come up with a decision that left the whole affair up in the air and no nearer a resolution. After a year of relative calm between TML and Eurotunnel, suddenly the rows were running again. And for the first time TML voiced what many had feared: that the way this project was going, it was not going to finish anywhere near on time.

Still gagged by the contract officially, TML was now adept at getting its message across through the press by the use of unnamed "sources close to the contractor". "I am not at all confident that this project is going to finish on time. Unless there is a considerable change in attitude by Eurotunnel, the scheme will come in late," said one source.

TML had no interest in the project finishing late, any more than Eurotunnel did. But it did know that a late completion would hurt Eurotunnel more than it would hurt the TML member contractors. More than anything, Eurotunnel – with the bankers on its back – needed trains running through a completed tunnel and money coming in to start paying off its debts. If threatening late opening was the only way to scare Eurotunnel, then TML, when provoked, would raise the subject. What was now obvious on site was that the programme was slipping and only close cooperation between the sides on the design of the fixed equipment and subsequently on the commissioning of the project could result in a chance of a timely opening.

All three tunnels might break through on schedule, but the disruption over the fixed equipment meant that work was almost at a standstill and the tunnels were standing virtually empty at a crucial time. TML blamed Eurotunnel for the disarray. Eurotunnel blamed TML. And the disputes panel was not helping at all.

The farce over the panel ruling and subsequent row had stung TML into voicing a barely veiled threat that if Eurotunnel did not stop meddling and instead get on with finalising its requirements and paying what it owed, then it could forget cooperation on site. And without co-operation, the tunnel would open late. It would be by no means the last time TML played this card during the bitter rows ahead.

14 Building the Channel Tunnel

"I used to enjoy taking people to the crossovers. I'd bundle them into the lift then pack them tightly into the passenger loco for the journey to the crossover cavern. Then I'd watch their faces as they emerged into the huge chamber."

Colin Kirkland, Chief Engineer, Eurotunnel

The actual construction of a 50 km-long Channel Tunnel up to 50 metres below the sea tested the ingenuity and skills of the top minds in the European and American construction industry. Although the tunnelling itself was not expected to require the use of any novel technology, the logistics of the whole operation were more complex than anyone imagined.

The terminals that were built at either end of the project were immense construction projects on their own. The infrastructure built around each terminal formed a major civil engineering contract of a size that would warm the heart of any work-starved major contractor. Between the terminals the three 50 km-long tunnels were built using 11 massive tunnel boring machines working on 12 separate tunnel faces. These tunnels were connected every 375 metres by 130 hand-dug cross-passages 3.3 metres in diameter and the two outer tunnels were joined every 250 metres by tunnels that formed 2 metre diameter piston relief ducts.

The three main tunnels ran into two vast undersea caverns 160 metres long, 11 metres high and 18 metres wide. The construction of these crossovers tested the nerve and skill of the finest engineers in the world.

In the UK the main site was at Shakespeare Cliff near Folkestone. From here the six British tunnelling machines started their journeys – three seaward to meet their French equivalents in the tunnel and three landward to break through 8 km away at the terminal site at Cheriton.

The Shakespeare Cliff site was used in the abandoned 1975 attempt and is arranged on two levels with the upper level housing offices and support services 65 metres above the base level by the sea. This lower level was built in 1843 during the development of the railway line from Folkestone to Dover.

When TML took over the site the two levels were already connected by a road tunnel built in the 1970s attempt. At the lower site was an access tunnel, or adit, running down to a 450 metre length of tunnel excavation left over from the previous attempt. TML sunk a 110 metre shaft from the top of the cliff down to these existing tunnels to provide access to the workface for materials, machines and the workforce. It also dug a second adit from the lower site, increasing access to the key area where the tunnelling machines would start their journeys. By enlarging the old stretch of tunnel at its seaward end enough space was created for the first tunnel boring machine, which drove the service tunnel, to be assembled underground after being lowered to the site down the access shaft.

This started its journey towards France in November 1987, now backed up by the site rail transportation system and the first tunnel linings, which were being stored on the lower site after delivery by train. High-speed gantry cranes were erected to handle these linings onto the delivery trains for transportation to the workface.

Using hand digging methods – which actually involves using standard excavation machinery while supporting the roof with anchors and concrete – a large chamber was created under the lower Shakespeare Cliff site, and this gave TML the room to assemble the other five tunnelling machines, which were due to leave the site over the next 12 months.

At the same time a sheet piling wall, one of the longest in the world, was being built in the sea around the lower site to enclose an extension to the site created by the disposal of millions of cubic metres of spoil arising from the tunnel excavation. As this platform grew it was occupied by batching plants and workshops. Despite some environmental objections to the creation of the platform, it did allow an easy way of getting rid of the spoil and provided extra space for the contractor. The alternative would have seen thousands of trucks driving through Kent to waste tips – if any could be found – to dump millions of cubic metres of chalk.

In France there was nothing as convenient as a lower working site to aid construction. Instead, the French sank what was dubbed the Great Shaft of Sangatte, from ground level down to a depth where tunnelling could begin. The shaft, at 75 metres deep and 55 metres in diameter, had to be big enough to allow the tunnel boring machines to be lowered into position to start their journeys.

Whereas on the UK side the 12 metre-long cutting heads of each machine had to be dismantled to allow installation into the tunnels, the French shaft was large enough to take them whole, lowered into place by cranes with a 430 tonne lifting capacity.

The French had an ingenious method of disposing of their tunnel spoil. At the base of the shaft, some 15 metres below the platform from which tunnelling began, the chalk spoil was mixed with water until it was capable of being pumped a mile away behind a dam for storage.

From these two points – Shakespeare Cliff and the Sangatte shaft – the tunnelling machines began their journeys.

The tunnel boring machines were about 250 metres long and comprised a circular boring face – the size of the external diameter of the tunnel – backed up by an ingenious collection of rams, lining erectors, conveyors and handling cranes. The principle adopted by the UK machines was relatively simple. The circular cutting face excavated spoil by rotating while being pushed forward by four hydraulic rams. The excavated spoil passed through holes in the face of the machine onto a conveyor, which carried it away along the excavated part of the tunnel and deposited it into wagons waiting to take it to the portal for disposal.

The cutting head was typically about 12 metres long and housed the main drive and bearings, a control centre with closed-circuit television and the laser guidance system. At the back were four main rams which, by pushing back against the completed concrete lining behind, forced the cutting head forward to apply the pressure that allowed the head's circular action to excavate spoil.

This made the concrete lining an integral part of the excavation process rather than just a supporting skin for the completed tunnel.

For the most part the lining comprised 380 mm thick concrete, but this varied according to ground conditions and was more than 500 mm thick in the under-land section. In the UK running tunnels each 1.5 metre-long lining ring was formed of eight segments plus a key segment, whereas in France six 1.4 to 1.6 metre-wide segments plus a key were used. These were cast outside the tunnel and delivered to the machine on the wagons. More than 700,000 of these segments had to be cast by TML, and with differing ground conditions demanding different linings, the segments could weigh anything from 0.75 to 9 tonnes. In France 250,000 segments were cast at Sangatte immediately above the huge shaft.

In the UK, of course, things were not that easy. With barely enough room to hold a concrete mixer at the base of Shakespeare Cliff, TML had to find an alternative location to cast the UK side's 450,000 segments. This was eventually found at the Isle of Grain in Kent, between the River Medway and the Thames Estuary. Here a remarkable segment casting production line was set up, producing segments in the strongest concrete the world has ever seen.

Concrete is measured by its crushing strength in newtons per square millimetre, achieved a certain number of days after it is cast. Ninety days after casting the concrete used to form the lining of the tunnel a force of

70 to 90 newtons per square millimetre was required to crush it. The figure for the concrete in the pressure dome of a nuclear power station is 50 newtons per square millimetre.

These segments were taken 60 miles by rail from the Isle of Grain to Shakespeare Cliff and stored on site ready for delivery to the working face. When they were needed the tunnel wagons took them to the boring machine, where they were loaded by a handling crane onto segment cars. The machines could carry enough segments for two full rings. These were then delivered along the machine to a storage area before being loaded into a magazine ready for installation.

The cutting face excavated 1.5 metres of tunnel – the length of the tunnel linings. Once this was done the rams were retracted, leaving a gap between the cutting head and the completed tunnel lining. At this point the segments were placed around the excavation by two cranes – a lower segment erector and an upper segment erector. The last segment placed was the keystone, a slightly smaller segment than the others. It was forced into place, forming a strong ring of concrete. The rams then pushed back on this new section of tunnel lining to repeat the process, while grout – a mixture of cement and water – was injected behind the lining to fill the gap between the earth and the concrete.

When working efficiently the system was excellent, as TML found in the heady days of 1990. But it relied on many factors to guarantee smooth operation. The production line from the segment casting areas must be set up and working well. The right segments must get to the tunnel machines at the right time, and the spoil wagons must quickly get the spoil away.

The other important factor was that the ground through which the machine was boring must be capable of supporting itself over a 1.5 metre stretch for some time. This was essential during the stage when the rams had pushed the cutting face forward by 1.5 metres and then retracted, leaving a circular section of excavation awaiting lining for up to an hour.

If TML had expected wet ground on the UK side then they might not have used this tunnelling technique. Indeed, on the French side, where it was well known that the ground was treacherous wet sand, the machines used were considerably different and in fact more sophisticated than those in the UK.

In France the cutting head could be operated in closed or open mode. In poor conditions closed mode was used. This isolated the cutting equipment from the back-up machinery and prevented water bursting into the tunnel behind the cutter. The machines used a complex arrangement of seals, which meant they could withstand up to 10 atmospheres of pressure.

The tunnel linings used in France were either cast iron – where the ground was particularly poor and where a good seal was required – or bolted concrete. Whereas in the UK the lining was fixed into place by expanding it using the keystone, the French used the bolted system to make life easier in the wetter ground.

The ingenuity of the tunnel machine manufacturers allowed the French TBMs to be adapted to open face working in better ground conditions. The closed tunnelling face discharged the spoil onto an Archimedes screw conveyor which transferred it to the back of the machine. The problem there, of course, was that in the high-pressure areas the force applied by the spoil from the front could push it back through the cutting head and into the back-up areas of the machine. A brilliant but simple system prevented this. A second Archimedes screw conveyor was fixed to the end of the first on a dogleg. This second screw turned slower than the first which meant that a plug of earth was constantly forming at the junction of the two, sufficient to absorb the energy generated by the pressure outside the machine.

Colin Kirkland was Eurotunnel's technical director at the time of the tunnelling and remembers with affection and admiration this stage of the work and how such problems were overcome. "The Archimedes screw system is a simple Japanese invention, but it works," he said. "And the beauty is that if you want to go to open mode all you have to do is bypass the second screw and you are away."

In open mode progress was always much faster, so in view of the French ground conditions a flexible machine was essential. But it was, of course, much more expensive than the machines ordered and built for the British teams.

Kirkland reckons that the project was just a bit unlucky with the ground conditions on the British side. All the pointers were that the machines would have a good clear run through the perfect tunnelling medium – good, dry, impermeable chalk – but within a few metres of starting tunnelling with a machine designed for working in the dry, the UK side hit the wet ground which almost brought tunnelling to a halt. "The 1970's tunnel had stopped in excellent dry chalk and investigations had indicated that would be the case in the early part of the work," said Kirkland. "But it turned out to be a lot wetter and more fissured than that and the machine struggled in the first weeks (in 1988)."

The wet ground would not play ball and remain unsupported over a 1.5 metre length while the lining was expanded into position. The chalk was microfissured, which allowed water to pour through into the machine, which in turn led to the chalk collapsing into the tunnel between the last lining ring and the protective shield of the cutting head. This caused huge disruption and great discomfort to the miners. The workers struggled to place each lining ring because the chalk moved, thereby reducing the diameter of the excavation. In short, the linings just would not fit into place and installing each one became a major operation. And of course the

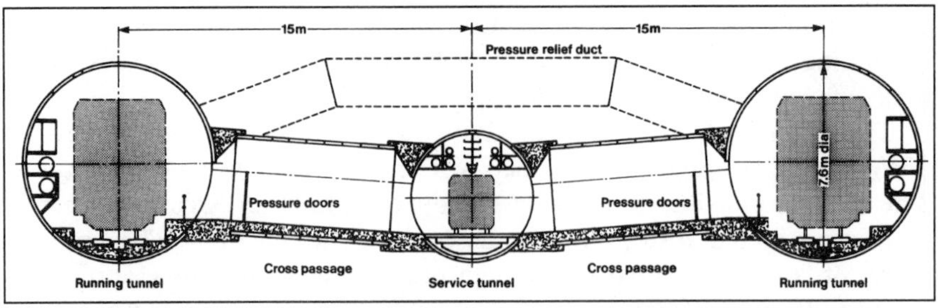

(Above) Cross section through the tunnels at the cross passages, and (right) through a running tunnel showing the mechanical and electrical services.

salt water poured into the machinery, damaging the hoses and electrics and compounding any problems in the tunnel.

"I reckon that if the bad ground had cropped up half way through the project it would not have caused so many problems," said Kirkland. "But coming as it did during the learning curve and in an unexpected area it made the delays all the more severe."

The upshot was that TML had to modify the machine in situ, 50 metres below the sea. A series of bars or "trailing fingers" were installed behind the cutting head, so spanning the gap between the head and the last section of lining. These bars held the chalk at bay, allowing the segments to be erected as soon as possible.

TML also took the precaution of applying an extensive waterproofing regime to the machinery and hoses to protect them from further salt-water attack. With these adjustments made, the TBM started to make better progress through the rough ground and using the concrete lining.

Expensive cast iron linings rather than concrete had been used earlier, and indeed in those early months more cast iron was used than the total expected for the whole project.

Two more tunnelling machines had yet to pass through this poor ground, of course. This first drive was only the central service tunnel. Yet to come were the two running tunnels – the north and south drives.

The machines used were considerably larger, with an external diameter of 8.36 metres compared with 5.36 metres. But TML was able to adapt these machines before tunnelling began in 1989, to allow for the poor ground which they were now all too well aware was ahead of them. But still further modifications also had to be made to the machines in 1989 and 1990.

The poor ground conditions were forcing TML to inject grout not just behind the tunnel linings, but also ahead into the soil they were about to excavate. This had the effect of stabilising the excavation and reducing the water ingress and chalk collapse. But it was an unexpected operation and

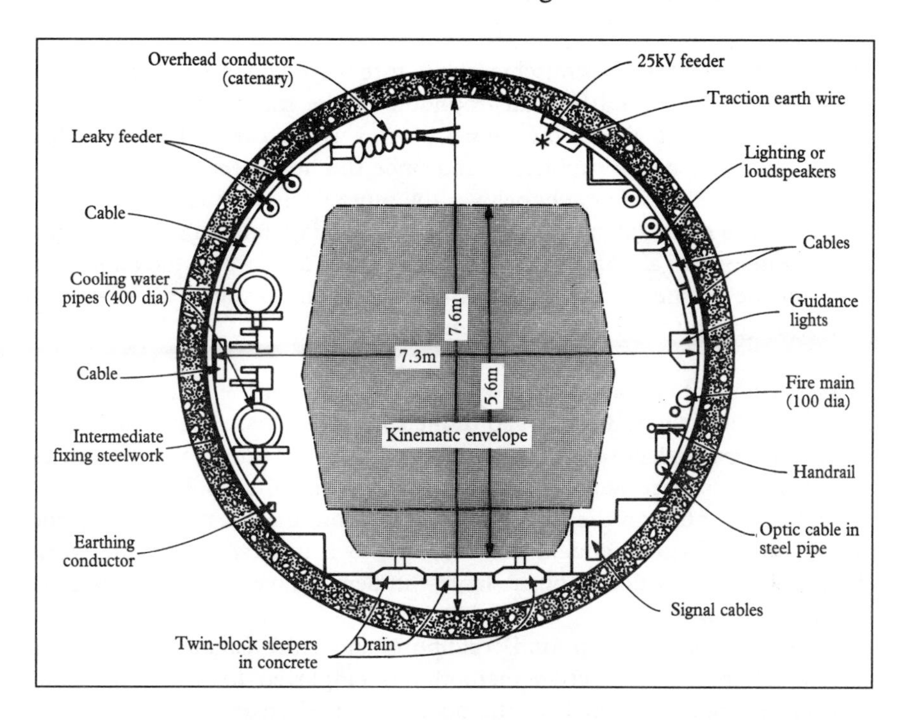

was being severely hampered by the segment delivery system which brought the lining pieces to the face along both the top and bottom of the machine.

By adjusting this so that segments were only delivered at the higher level – a technique that sounds simple but was made outrageously complex underground – grouting could go ahead at high speed.

It was a combination of these modifications, plus grout stabilisation of the ground ahead of the running tunnel machines by injection from the service tunnel, which allowed the machines to move forward and through the wet chalk. A subsequent improvement in ground conditions resulted in rapidly improving progress in tunnelling in 1989, until world records for tunnelling speeds were broken in 1990.

If the timing of the poor ground was bad – coming when the tunnellers were learning their routines anyway – then at least it had the effect that when good ground was reached the underground teams found life remarkably easy. No water, no collapses and they were away, picking up pace all the time towards a breakthrough that incredibly – considering the early problems – was on schedule.

The service tunnels played an important part in the construction of the two mammoth crossover caverns that were carved out of the chalk to allow trains to switch tracks. Eurotunnel's technical director Colin Kirkland, who conducted countless tours of the site, enjoyed showing off the crossovers to visitors. His mischievous nature played on the concerns of his guests that they would be claustrophobic in the tunnel. What Kirkland knew was that the size of the tunnel bores was such that claustrophobia would certainly be no problem – and once the huge crossover chambers were complete then agoraphobia was the more likely complaint.

"I used to enjoy taking people to the crossovers," remembers Kirkland. "I'd bundle them all into the lift then pack them tightly into the passenger loco for the journey to the crossover cavern. Then I'd watch their faces as they emerged into the huge chamber."

Two permanent crossover chambers were built under the sea. Their function is to allow the trains to swap tunnels for maintenance and safety reasons. The crossovers split the tunnels into three sections so that maintenance can be carried out in one area without closing the whole tunnel. One chamber was built on the British side, 7 km from the coast, and the other 12 km from France, and in the case of the British crossover the work involved the construction of the largest undersea cavern in soft rock anywhere in the world.

The variation in the ground conditions under the Channel demanded that a different construction method was employed for each, with the French technique adapted for the poorer fissured chalk prevalent off the coast at Calais.

Construction of the British crossover began first in June 1989.

The chamber had to be an immense 160 metres long, 18 metres wide and 11 metres high, allowing the two running tunnels to enter and the tracks to cross to allow trains to switch from one tunnel to the other if required. However, the design of the Channel Tunnel away from the crossovers demands that the service tunnel – the smaller-diameter bore – be driven in the middle between the two running tunnels. This clearly could not be the case at the crossovers, so the first action in the crossover construction sequence was the diversion of the service tunnel off to one side to dip below the line of the northern running tunnel.

"The service tunnel was bored ahead of the two running tunnels," said Kirkland. "So once it had reached the site of the crossover it changed course before again running parallel with the original line. The crossover was then built from adits dug from the service tunnel."

The technique chosen to build the chamber was the New Austrian Tunnelling Method. This involved excavating sections by hand and supporting the roof with rock anchors and shotcrete – a sprayed concrete. The amount of support required was determined by the movement in the chalk, which was constantly monitored by the engineers. Normally NATM is used in hard rock, but the Channel Tunnel saw its first UK application in soft rock. On a chamber the size of the crossover, this was an extremely brave move.

"We chose to build the crossover by excavating different sections at a time," said Kirkland. "So having dug access adits from the service tunnel to the correct location underground we then began digging two small tunnels, one either side of the crossover location."

These tunnels essentially allowed the side walls of the chamber to be cast in concrete before the excavation of its main body. Initially hand excavation was used, but as soon as plant could get in then it was installed to get excavation fully under way.

"The plant was delivered in pieces and assembled underground in the crossover chamber once the first section had been hand excavated," said Kirkland. "It all had to be delivered to the right location on special trolleys that spanned both tracks in the service tunnel, so the deliveries were made during a weekend closure of the service tunnel. You have to bear in mind that while the crossover excavation was under way, the service tunnel was still carrying on and the spoil from that still coming out and materials going in. In the end it all went extremely well."

Once the two side tunnels had been dug, lined and the chamber wall cast in concrete, then excavation began of the top section of the crossover. During this stage the roof of the chamber was supported by the shotcrete spray lining, with rock bolts installed where required and as frequently as demanded by the movement of the clay above.

"The whole construction of the chamber was a courageous step," said Kirkland. "But at this point we were forming the arch 35 metres below the English Channel while monitoring the movement of the clay above. We were happy with a movement of 20 mm or so and even 40 mm would be OK, but around 80 mm would be worrying."

On the whole the movement was as predicted, but there were some hairy moments for the tunnellers and in some areas the density of rock bolting was quite high.

With the top section successfully removed, out came the bottom and the shotcrete-lined chamber was complete. In August 1990 the first running tunnel machine broke into the crossover cavern, having completed its 7 km drive from the coast. It was then moved across the chamber and restarted its journey towards France at the far end.

Once the second running tunnel machine had also gone through the chamber then the invert was excavated and the chamber's permanent concrete lining – 1 metre thick at the top and 1.2 metres thick at the bottom – was cast, completing one of the most remarkable civil engineering achievements of the modern era.

On the French side the ambitious construction of a large open arch 35 metres below the sea bed was considered too risky for the construction of their crossover. The ground conditions on the French side were expected to be that bit more fractured and fissured than on the British side, so the fear of collapse or excessive movement of the roof was too great. The solution to the conundrum was to use a variation on the multiple drift excavation method used on the British side.

"Basically what we did was excavate the French crossover in smaller bites," said Kirkland. "The running tunnels drove through first, as opposed to the British method, and the chamber was then formed around those two main bores." A series of small tunnels were then bored in an arch around the main drives. These mini tunnels were then filled with concrete to form the arch, allowing excavation of the chamber below the new roof.

Although this method was successful the big disadvantage was that it had to be built late in the construction programme, once the running tunnels had passed the crossover site 12 km away from the French coast. Construction of the second crossover was completed in November 1991, almost a year after the first tunnel breakthrough.

While all this underground activity continued, there was still plenty to do above ground. On the UK side the tunnels emerged 900 metres short of the terminal and three different tunnelling methods were required to complete the journey.

The New Austrian Tunnelling Method – similar to that used for the crossovers – took the route through the geologically challenging Castle Hill, while either side of the hill cut and cover construction and top-down construction were used. Cut and cover work involved excavating the area and building the tunnel out of reinforced concrete boxes. Top-down construction – used where space was tight – involved building the roof of the tunnel first and then excavating below it. In just a 1 km stretch the project demanded four different methods of tunnelling – a reflection of the challenge that some of the less publicised sections of the scheme could throw up.

The UK terminal was a major site in itself, and at its peak was the third biggest construction site in the UK behind Canary Wharf in London's Docklands and the Sizewell B nuclear power station in Suffolk. The site had been stabilised to prevent nearby hills causing slippage in the area and the site level raised by up to 12 metres, with 2.5 million cubic metres of sand dredged from the sea and delivered by pipeline.

The terminal required 17 bridges to be built and a loop tunnel formed by the cut and cover method. Four of the bridges are huge – up to 350 metre-long concrete structures – and from these the ramps descend to take the cars to the shuttle trains. Add to this the many buildings and services on the 2.5 km by 900 metre site, and you have a construction scheme of major proportions.

But the French terminal is five times bigger than the one in the UK and covers a greater area than Heathrow Airport. A huge earth stabilisation programme was carried out to turn marsh at Coquelles into a solid foundation for the terminal and a complex canal and elevated water tank system was built to drain the area.

TML had to excavate up to 30 metres deep to form a slope down to the tunnel entrance, and build a complex of bridges, ramps and viaducts to take traffic on and off the shuttles, as in the UK.

The amount of construction required for the scheme in addition to the tunnelling was immense, and for five years TML's operations dominated the economies of the north of France and south of England. The magnitude of the work ancillary to the tunnels underlined the fact that the Channel Tunnel was as big a civil engineering venture as anyone could remember.

15 King of the tunnels

"I've never worked harder than I did in those days. It was the greatest thing I ever did and the greatest project I worked on."

John King, Tunnelling Director, TML

In the autumn of 1989 TML Tunnelling Director John King stood in the UK's south marine running tunnel and watched in horror as water poured into the excavation and chunks of chalk up to 1 metre deep collapsed onto the boring machine. It was, said King, the low point of the whole project for him. "I stood in the machine with the tunnel superintendent and the water was just flooding in. We knew from the service tunnel excavation that this could happen – but it was on a much larger scale than we expected. We watched the water come in and tried to work out what on earth we were going to do."

John King was TML's tunnel guru. He was brought onto the project in June 1988 to sort out the awful problems that the contractor had got itself into on the service tunnel work. King had plenty of experience – he knew tunnels and tunnellers better than anyone in the United Kingdom. He had worked for contractor Mowlem for 40 years and become an expert in tunnelling and foundations. His successes included the construction of underground GPO tunnels in London, the Piccadilly and Victoria Line extensions of the tube and Newcastle's Tyne and Wear Metro. In 1986 he left Mowlem to become a consultant to London Underground.

So when, after six months on site, TML realised it needed a new approach to get the tunnelling work out of trouble, King – at 62 years old – was the man the contractor called in to save the day.

King became the tunnel hero. When he started the work was in chaos, the morale was low, water was a hazard and the work was already six months late. When he left all the tunnels had broken through on schedule.

"I was invited in to strengthen TML's management and take on the technical and ground problems. I sized up the problems as soon as possible – and what was clear was morale was terribly low. So I made some management changes and brought in some engineering managers to run the site."

The service tunnel machine was struggling through the poor ground at that time and moving forward at a rate of about 50 metres a week. With 20 km or so to drive, this was a major worry to TML and its team. "The problem was trying to fit the expanded lining system chosen for the job in ground that was falling away," said King. The concrete lining was fixed into place by forcing home a keystone. In France, where poor conditions were expected, the concrete segments were bolted rather than forced into place.

But with thousands of concrete lining segments already cast in the UK there was no option of changing the system. TML had planned to use a cast iron lining in poor ground, but this was hugely expensive and with the unexpected ground TML had to find a way of fitting the concrete lining into position.

"We had changed from concrete to cast iron but it was very expensive and much slower," said King. "The problem was that with this poor ground being unexpected then no-one had any idea how long it would last. We had to assume the worst and set up systems to cope with it. If we knew that it was coming we would not have used this tunnelling method."

King got the tunnelling team and the TBM manufacturer Howden to work closely together and develop modifications to the machine – the trailing support fingers described previously – that would allow the expanded lining to be fixed. "It was a double expansion method," said King. "The linings were fitted while the trailing fingers held the ground in place and partially expanded. Then the TBM moved forward, preventing the trailing parts being trapped behind the lining and allowing the lining to be fully expanded into place. The team developed this underground – it was very innovative."

The other machines that were due to follow behind the service tunnel machine were similarly modified, but as they started their journeys in 1989 the water problems posed in the larger-diameter tunnels were considerably worse. "We realised that we still had not solved the problem," said King. "The water was coming in and these large chunks of chalk were falling into the tunnel. It was dangerous, time consuming and very wet."

The tunnel team's solution was to boost the grouting-up operation behind the machines. Previously the grouters were working 100 metres behind the tunnelling face. But by modifying the machinery in the TBM the grouters were brought forward to nearer the face to stop the water affecting the work. Also, grout was injected over the top and ahead of the running tunnel drives by grouters working in the service tunnel. "This brought us an element of watertightness and the machines started progressing better. But I knew there was another major problem – the transportation system. The electric battery locomotives which were meant to haul the materials in and the spoil away were struggling. They could not cope with the wet and we had a long saga getting these right. There were major modifications made to the locos to stop them packing up in the salt-laden moist atmosphere and to boost their power and strength. We had to make them heavier to boost the friction with the slippery rails and waterproof every part of the machines."

King then insisted on setting up a separate organisation within TML to run the transportation system, a move that was unpopular on site. "We eventually had the third largest underground railway in the UK just doing the deliveries and disposal. I decided it had to be run by railway people, so I brought in a manager from British Rail to run the team and created a separate organisation to run the system safely and efficiently. My tunnelling colleagues did not want this at all and I was pretty unpopular on site for a while. But the logistics of this project were more challenging than the tunnelling and I think my system proved itself when the scheme got to the complexities of the fixed equipment installation that followed."

King was particularly pleased when, with the crossover and the cross-passages being built, the system still coped with delivering and removing material to the various sites.

King was involved in the drive to improve safety on the site and looks back on the poor 18 months between January 1989 and July 1990 with regret. "I've thought back to the problems we had then many times. We realised in 1989 that we needed a completely different approach to the health and safety on site. We went for a complete culture change and eventually achieved it. A huge amount of training was required and Jack Lemley introduced the Dupont safety system, which changed things. This introduced a culture of talking safety constantly – we put safety in everyone's minds all the time and we found you could break tunnelling world records safely. Our principal motivation was a moral one – you cannot be a successful manager and have an inadequate safety record. In the end our culture change worked."

King looks back with huge affection at this project and says the highlights were the breakthrough and the completion of the crossover chamber. "Nobody had built anything like this using the New Austrian Tunnelling Method in soft rock before. That was a great achievement – really great engineering. What struck me about the Channel Tunnel was that it was bigger than anyone had dealt with before and bigger than anyone could imagine. I've never worked harder than I did in those days. It was the greatest thing I ever did and the greatest project I worked on. It was a great end to a management career."

King left the project at the end of 1990 after the first breakthrough had underlined his success. But he stressed it was a team effort. "I had a remarkably hard working and terrific team. While Eurotunnel and TML were bickering at the top level, on site we were getting on with it and building the tunnel."

It may have been a team effort. But it was a team built by John King.

16 TML's £1.4 billion bill

"We are running out of time. We must settle our problems right now if we are going to save the project."

Jean-Paul Parayre, co-chairman of TML

In the summer of 1990, before the first tunnel broke through under the sea, Eurotunnel's project chief executive John Neerhout remarked that he thought people's perception of the project was all wrong. "It surprises me," he said, "that people still view this as a tunnelling job. It's not – it's a transportation project."

The point Neerhout was making was that, come breakthrough that winter, the job would be far from finished. In many ways it was just starting. The tunnels are merely the vessels to carry what will be one of the world's busiest railways. A complex state of the art railway system would have to be squeezed into these tunnels and then tested, commissioned and operated. It would make the tunnelling look like a simple enabling project carried out in advance of the real work.

The tunnelling, though, was romantic. It caught the public's imagination unlike any construction project. The tunnels joined Britain and France for the first time since the Ice Age and joined the island to the continent.

The tabloids glorified the "tunnel tigers", some of whom, they said, battled with appalling conditions in 12-hour shifts to carve out the tunnels before spending some of their £1000 a week pay packet on immense quantities of alcohol and terrorising the female population of Kent.

When breakthrough came in December 1990 everyone knew about it. The faces of Graham Fagg and Phillippe Cozette – the two tunnellers who shook hands through the excavated hole – were plastered over every newspaper and television channel. To the public the ceremony gave the impression of a project nearly complete. You could almost hear the sounds of parents hitching up the four-berth caravan and revving up the Ford Sierra ready for the first trip on a train to France. On site the mood was rather different.

Joe Stacey, a senior engineer at Eurotunnel working on the mechanical and electrical side, said at the time: "What we must do is turn this structure into a fully working railway ready to carry passengers in 1993." He knew at the time that the task ahead was immense. The work would have been complicated enough if the planning had been perfect. But the planning had not been perfect. Far from it.

A year before the breakthrough none of the major M and E contracts had been let. Even when they were awarded TML struggled to get its programme in order because on the day the tunnels broke through it still did not know exactly what Eurotunnel wanted.

There were several reasons for this. Poor planning early on in the contract, changes to the specification forced by the Channel Tunnel Safety Authority, disputes about the quality of specification required and promised for the tunnel, and attempts by both sides to rationalise the fixed equipment to bring down costs. The result was chaos on site just at the time when organisation and efficiency were the keys to successful completion.

Three 50 km-long tunnels had to be converted into a working railway by the installation of catenary systems, cooling pipes, drainage, tracks, signals, fire systems, lighting, communication networks...the lot.

The only way in was through each end of each tunnel, of course, so the problem of getting the equipment to the correct part of the site was huge.

In all, 550 km of pipe for the drainage, fire and cooling systems requiring 120,000 supports had to be installed. For the electrical system an incredible 1300 km of cable had to be fitted on 350 km of cable trays, all bolted to the tunnel walls, and the tunnels were to be lit by 20,000 light fittings. More than 600 doors – massive metal structures rather than convenient 2 metre-high wood-panelled affairs – had to be fixed. And once in place everything had to work.

The whole operation was a logistical nightmare, with TML looking to manage more than 40 subcontractors all vying for their place on the tunnel delivery trains.

The right equipment had to arrive at the right place at the right time and in the right order. The delivery trains could take more than an hour each way to reach the more remote areas of the tunnel, and a forgotten item could see workers twiddling their thumbs for vital hours.

In the first few months of installation there were two additional serious obstacles which slowed work. First, spoil was still being excavated from the tunnels when fixed equipment installation began, so the number of locomotives running underground was high. Not only did the locos have to drag equipment in and out of the excavation, but spoil still had to be removed and concrete linings brought in.

Spoil wagons had the right of way in the tunnel because they were so heavy and cumbersome that a loss of momentum caused serious difficulties for the locomotives pulling them. TML's locomotive schedule for deliveries had to take this into account, further complicating a difficult timetable.

The second problem in the early months of fixed equipment installation was even more disruptive to the loco schedules. The temporary track that had been laid in the tunnels on which the construction transportation system ran had to be replaced with the permanent track. And the temporary lighting and ventilation system which had enabled excavation to proceed also had to be removed and permanent systems installed.

The temporary track was narrow gauge while the permanent way was standard gauge, so the same trains could not run on both type of track. Deliveries had to be planned to match the correct type of system for the track available in the tunnel. In the end, there was no track at all in the service tunnel and deliveries were carried by wheeled vehicles, while in the running tunnels only one track was available for construction where previously there had been two.

TML had little choice but to run the work like a production line, with a materials controller – brought onto the job from the motor industry – given the lead role in the coordination and planning. His job was to follow the process of procurement, ordering, delivery to the tunnel site, delivery into the tunnel and fitting of each element of the works. The problem was that if one item was not ready at the right time it could hold up the whole process and slow work to a crawl.

TML eased the difficulties by building four diagonal cross-tunnels connecting the three main bores and allowing delivery locomotives to switch between the three during the services installation phase.

TML and Eurotunnel were both well aware of the difficulties ahead in early 1991, particularly as TML had already got off to a bad start with teething troubles in the mechanical and electrical fit-out phase of the work.

Some of the original suppliers of primary supports and brackets had to be replaced because of troubles with scheduling, and some of the work carried out in the early weeks had to be redone – wasted time that TML could ill afford.

By the beginning of 1991 the delay on site was put at about six months and progress was still painfully slow. TML still laid many of the difficulties at Eurotunnel's door because of the lack of information on exactly what Eurotunnel wanted in the tunnel.

With the problems on site mounting, Eurotunnel was hit by yet another blow.

Sir Alastair Morton – knighted in the 1991 New Year's honours list which also bestowed an honorary knighthood on co-chairman André Bénard – must have hoped that one day he could attend the twice-yearly press conferences which accompanied Eurotunnel's interim and final financial results with a bundle of good news for the media.

A couple of times he had managed to dress up news "not as bad as expected" as good news. But never had he been able actually to present a press conference where he could be sure that the next day's headlines would guarantee a share price rise for his company.

April 1991 was no different.

Although it was a low-key announcement by Eurotunnel's standards, the April report revealed a whole new problem for the developer.

The trains were not going to be ready for the tunnel on time.

As if it was not enough having to battle with a contractor over claims and worry over the progress on site, now the rolling stock that would go through the tunnel was destined to miss its delivery date.

Although Morton said the tunnel would still open in June 1993, on that date only a limited tourist shuttle service would start and build up to a full service only by December 1993. The delays in the shuttles were due to increased safety features being demanded by the Intergovernmental Channel Tunnel Safety Authority, which had forced Eurotunnel to change the size of fire doors on the trains.

The safety authority was charged with awarding a licence to the tunnel to operate and so scrutinised every aspect of the work to ensure it would be safe to run. Its decisions on the shuttle trains were to have a marked impact on the scheme, and indeed result in some animated discussions before compromise designs were agreed.

In February 1991 the authority had ruled that the doors through the fire barriers between the passenger wagons should be widened by 100 mm. In response, the consortium building the wagons – Bombardier of Canada, BN of Belgium and ANF of France – said the shuttles would be four to five months late, so the full quota would not be ready in time to run a full service. The safety authority at the same time refused to accept Eurotunnel's designs for open-sided HGV shuttles, as they feared that if a truck caught fire, the air flow through the sides would fan the flames and the blaze would become uncontrollable.

Eurotunnel would spend many months arguing its case for open wagons with the Authority, because the cost of closing in the trains already under construction would have been astronomical.

The authority's rulings angered Morton. Not necessarily because he thought they were wrong, but because the decisions were made so late that work was already under way and the disruption caused by any design changes was multiplied. He took time away from his favourite hobby of slamming the contractor to criticise the governments for being slow to handle the safety issues, which were jeopardising the programme on site and the manufacture of the trains. It was a theme taken on towards the end of the scheme by project chief executive John Neerhout when he looked back at the lessons that should be learnt from the scheme. "The first thing is to have the scope of the project well understood from the beginning. For instance, the Intergovernmental Commission (which created the Channel Tunnel Safety Authority) did not get organised until the job was well under way. The delay may well have been for good reasons but in my view it would have been better not to start until all these essentials were in place." It now seemed that the lack of preparation was going to cost the project dear.

In Eurotunnel's annual report – published just after the announcement of the shuttle delays – the site problems were spelt out for the first time. The terminal traffic management system was revealed as being some 10 months behind schedule and the control centres three and a half months behind. The report also said that TML's claim had now risen to £1.112 billion from £953 million.

During the summer the situation on site did not improve greatly and in September TML moved to put things right in the now traditional Channel Tunnel method – by bringing in a top American contractor.

Keith Price, a main board director with Morrison-Knudsen, replaced Jacques Thibonnier as TML's managing director responsible for transportation systems and engineering. The move came immediately after the resignation of chairman Phillippe Essig and surprised Eurotunnel, who had been happy with the work Thibonnier was doing in very difficult circumstances. TML denied that either man's departure was connected with the delays building up on site, but said that Price was an expert in on-site matters. One TML source said: "The move is all to do with a change from the planning and procurement stage to the hands-on stage. M Thibonnier was an expert in planning and procurement. Keith Price is used to the out-of-office on-site problems, dealing with hairy-arsed builders."

By now independent observers of the project and sources at TML were privately admitting that, no matter how talented Price was, he was going to be unable to turn the work around sufficiently to see even a limited service open in June 1993. Materials deliveries were not coming through to the site quickly enough because they had not been ordered early enough, and it seemed only a matter of time before the developer would be forced to admit that opening in June 1993 was just too ambitious.

One of the main problems was that the commissioning of the project was expected to take six months after completion, and privately the contractors could not see site completion until March 1993 at the earliest.

Eurotunnel dismissed such concerns as TML propaganda put around to pressurise Eurotunnel to agree the outstanding claims. But seasoned tunnel watchers were now convinced a further delay was inevitable, and that Eurotunnel was taking a "head in the sand" attitude to the problems developing on site.

The project's problems were starting to affect the accounts of the ten member contractors, who had to decide whether to make provisions against possible losses. The five British contractors were keen not to do so. They said that if the claims were paid then there would be no losses. But the five French contractors took another view.

In France it is more common to take a pessimistic view on losses and make provisions immediately to avoid the tax burden. In Britain provisions are played down where possible in order not to scare the stock market. The French contractors in 1991 were keen to make provisions against losses on the tunnel. The British were not. The British contractors felt that the French had lost them ground in the claims war against Eurotunnel by effectively admitting that they did not expect to be paid. But the French position was that although some of the claim would not be paid, some would. This was a realistic assessment of a claim on a construction contract.

Contractors tend to claim more than they expect to get and it is rare, if not unheard of, for a contractor to win its full claim.

The first splits between the member companies of TML were beginning to show, splits which Eurotunnel was watching with great interest.

But by October 1991 TML was reunited in its animosity against Morton, who once again criticised the contractor's performance on site. In a letter to Eurotunnel's shareholders, in which he broke the news that the outturn cost of the project had risen again and was now £8.05 billion, Morton unleashed another broadside. He blamed TML for the loss of time on site and for the way that it was pursuing its claim for more money for the fixed equipment. "TML cocked up the early part of the tunnelling and recovered and we look for a similar recovery on the mechanical and electrical side," said Morton. He went on to hit at TML's management of the job, criticising the contractor for appointing 42 separate subcontractors to do fixed equipment work.

The letter also revealed that TML had changed its tactics on its massive fixed equipment claim. Instead of attempting to compile a detailed item by item substantiation of claims, as Eurotunnel had expected, in July the contractor had hit Eurotunnel with what it called a "global" claim. This claim said that the fixed equipment would cost £1.27 billion – as opposed to the £620 million original total – and that Eurotunnel should pay this sum plus a management fee of £160 million. This contradicted the terms of the lump sum contract, which called for claims to be broken down item by item, but TML believed the contract had been changed fundamentally.

TML felt that Eurotunnel had effectively ripped up the lump sum contract by making so many changes to the work that its very nature had been altered. Because of this the contractor was seeking a different way of being paid. The cost-plus management fee tactic was the contractor's solution – but one that Eurotunnel did not like or accept.

John Neerhout – as soon as he received the global claim – immediately called on TML to back its case for such a claim legally and to substantiate it as required under the contract.

But this TML could not do. An army of quantity surveyors working all hours would not have been able to put together the detail of every item in what was one of the biggest claims the construction industry had ever seen.

With no response from TML, Eurotunnel asked the dispute panel to rule whether it even had to consider the global claim. The Maître d'Oeuvre, Atkins-SETEC, agreed with Eurotunnel that the claim could still be settled under the original lump sum contract and that the global claim was not the way forward. But the MdO still acknowledged that the delays did entitle TML to make some sort of claim.

Morton's letter to Eurotunnel's shareholders also revealed that the MdO and the technical advisor were estimating a September 1993 opening date, while Eurotunnel was maintaining that a phased opening could begin in June 1993. "Progress is now dependent on the will and ability of TML's shareholders to organise TML to complete the project," the letter said. With this effort – backed by an earlier panel ruling that TML must do its utmost to finish as fast as possible – the project could still open on time, claimed Eurotunnel.

Unfortunately that cooperation and effort was difficult to guarantee. The claims dispute was starting to pinch. As Eurotunnel was refusing even to look at the new global claim, it certainly was not going to pay TML extra money for the fixed equipment work. This meant that each month the amount TML was spending was greater than the amount it was receiving from Eurotunnel. Although TML had cash reserves these were dwindling and the concern was that the 10 contractors would soon have to start putting their own money into the project to keep construction going. In this case TML would then be funding the project, not Eurotunnel, and immense pressure would build on the member contractors, who were already finding the going tough in what was now a fully fledged domestic construction recession.

The tension between the two sides created by the claim was so great that it was impossible to talk in terms of full site cooperation and a willingness to "mobilise wholeheartedly towards a June 1993" opening, as Eurotunnel described it.

TML was keen to point out that it had already accelerated its programme in an attempt to hit its 21 bonus milestones – points in the work it had to reach by certain dates to earn extra money – and that this acceleration was expensive. But without extra pay from Eurotunnel, TML said, it was not able to maintain the faster programme indefinitely.

Following Eurotunnel's letter to its shareholders and its financial report, which offered no crumb of comfort to TML, the contractor's leaders met in Paris to discuss their way forward. It was a meeting in which some of TML's members began to show their frustration at what they perceived to be Eurotunnel's intransigence. It was not so much that the claim had not been settled, but that Eurotunnel refused even to talk seriously about it.

The TML chiefs felt that their claim was genuine and the reason for not itemising the many elements was that such a course of action was impossible. They were not, they stressed, trying to hide anything or pull a fast one.

Eurotunnel, for its part, was able to delay payment quite easily by disputing the mechanism of the claim before even considering the validity of the actual case being put up by TML.

As experienced contractors TML's 10 member companies were well used to battling in claims disputes with difficult clients. What they were not used to, and did not care for at all, was a client that would not take a claim seriously. Worse still, it was a claim with considerable merit.

The contractor believed that Eurotunnel was becoming increasingly concerned about the rolling stock hold-ups and the possible costs to the project in revenue shortfalls if the start of the service was delayed further. The passenger shuttle delays were stretching towards six months and the HGV shuttles were now delayed indefinitely, as Eurotunnel tried to convince the safety authority to accept a semi-open wagon as a compromise to satisfy fire controls. A TML executive told *Construction News*: "There is a real risk that Eurotunnel could be faced with a tunnel that is completed but cannot be used because not enough trains are available for testing and commissioning the system. We are not the problem on this project – the rolling stock is."

In a perverse way this strengthened Eurotunnel's hand against TML. The client had to stop costs from rising any further, so had to take a hardline with TML. But any delay in TML's work, although extremely awkward, could well become irrelevant if the rolling stock situation did not improve.

TML's emergency Paris meeting discussed all these issues but inevitably returned to a key issue – that TML's funds were fast being eroded, at the rate of tens of millions of pounds a month, because of the income shortfall from Eurotunnel. The hawks at TML wanted to take some action immediately against Eurotunnel, and still wanted Morton removed. The doves tried to calm the situation, with the result that a compromise of a shot across Eurotunnel's bow was deemed most appropriate.

In a statement that followed the meeting TML said that its 10 member companies were "convinced that they have a strong contractual case for further payments. It remains the objective of all 10 shareholder companies of TML that their final revenue will at least cover costs and the shareholders have taken, and will take, all necessary steps to protect their interests."

This was seen as a threat to stop or slow work, and TML made it clear privately that this course of action would be taken if necessary and stressed that at the rate things were going the project was not going to open until March 1994. Selective action would be taken in those areas of work for which it was not being paid.

Indeed, at the same time as the statement was being issued TML was informing Eurotunnel that it would stop work on the cooling system if it was not paid extra money.

TML targeted the cooling system for very good reasons. Apart from being a key part of the scheme and on the project's critical path, the cooling system work had been totally redesigned. The original system had been estimated to cost £80 million but the new system which replaced it would cost £130 million. Eurotunnel would struggle to argue that this was part of the original contract and that if TML wanted paying extra it must justify its claim item by item, pound for pound.

To TML the cooling system was a prime example of the sort of work that was causing the difficulties. It had been totally redesigned, was clearly more expensive and, the contractor said, was outside the original contract. Pay now or the work stops TML concluded.

Eurotunnel acted swiftly. Rather than wait for TML to go to the courts to settle the claim, Eurotunnel decided to beat them to the mark. The developer applied for an injunction against its contractor, preventing it from stopping work on site, and threw in another attack on TML for good measure. "The contractors are clearly feeling under pressure and are throwing a lot of chaff to deflect comment and public attention from their own shortcomings," said Eurotunnel. "We continue to wait for TML to substantiate its claims and expect them to follow the contract which they wrote and to deliver the project on time as promised – which we are confident they are capable of doing."

TML immediately applied to stay the proceedings and the proceedings were adjourned while both sides amassed further evidence.

In most projects this would have been the nadir. On the Channel Tunnel there were so many low points that this was just another in a line of rows and crises which once again threatened to bring work to a disastrous halt.

With all three tunnels complete, it seemed impossible that the job would not now be finished and the link completed. But the depth of ill feeling between the two sides was now such that no-one could be sure of anything.

The hold-up in the court proceedings left TML still building the tunnel, still fitting the cooling system and still feeling mighty aggrieved. It had been three years since the five French contractors had broken the contract clause on silence. But the feeling was that it was about time for TML to do the same again.

TML was concerned that Eurotunnel was getting the ear of the national press and that the contractor's case was not being put across well, leaving the impression that Eurotunnel was battling against greedy contractors to hold on to what was rightfully the property of its banks and shareholders. Morton's colourful language and acerbic style made excellent copy for the newspapers and he carried on a love/hate relationship with journalists that guaranteed plenty of coverage for his point of view.

Indeed, Morton never underestimated the press or its usefulness and was diligent in his reading of the many column inches of coverage given to the project. And if he did not like what was written, then the editor and the journalist soon knew. Morton was a prolific writer of letters to various publications expressing outrage at any perceived errors in the copy or suggestion that the project might open late or need more cash.

TML, on the other hand, struggled to have its voice heard consistently. Unable to comment publicly the contractor relied on feeding information through to known and trusted journalists, but this was a hit and miss process that for much of the time had no coordination or control.

At the risk of being in breach of contract, the 10 TML companies announced a press conference in Paris that would "set the record straight with contractors' employees, shareholders and with the general public." TML was offended by the Eurotunnel letter to shareholders' which had challenged its professional reputation, and reckoned it was time to put its side of the story.

The conference was held on October 24 1991 and attracted print and television journalists from around the world. British and French journalists from the national and trade press crammed into the hotel vying for space with colleagues from America and Japan, many of whose banks were anxiously watching events and whose concerns had not been assuaged, even by personal visits by Morton.

All 10 TML member companies sent senior executives to Paris for the meeting. Jean-Paul Parayre, co-chairman of TML and chairman of Dumez, took centre stage alongside Peter Costain, his fellow co-chairman and chief executive of Costain. Neville Simms of Tarmac, who was looking after the technical side of the project, and Georges de Buffevent of Spie Batignolles, who held TML's purse strings, were the other speakers.

Tony Palmer of Taylor Woodrow, Sir Clifford Chetwood (Wimpey), Sir Robert Davidson (Balfour Beatty), Martin Bouygues (Bouygues), Jean Claude Jammes (SAE) and Antoine Zacharias (SGE) were also on the stage but refused to speak. The talking was to be left to the four appointed men.

TML wanted to ensure that they showed a united front to the public and had agreed on some basic rules among themselves before talking to the press. It was important, they felt, to hide any differences that the contractors might have between themselves over whether to make provisions in their accounts against tunnel losses and how far to go on criticising Eurotunnel and releasing sensitive information. By restricting the spokesmen to four they exercised a degree of control. And even when one journalist, perhaps surprised by the unusual silence from normally outspoken captains of industry, addressed Wimpey's Sir Clifford Chetwood directly, Sir Clifford dumbly shook his head as if to say anything at all would precipitate a collapse of TML's united front. If this was unusual, the opening of the conference was even more unorthodox. The first to speak was Peter Costain. The world's media settled down and waited with bated breath for the first words. Would Costain call for Morton's resignation? Would he announce that work on site was to stop? No. With the British and French trying to bury centuries of antagonism and mutual suspicion, Costain started by gloating over the English rugby team's controversial victory over France in Paris four days earlier. The British journalists winced. Their French colleagues revealed the full array of Gallic stares and shrugs in an atmosphere of stunned silence

But the difficult moment passed and TML settled down to tell the press, through Jean-Paul Parayre, that there were four conditions that needed to be satisfied if the Channel Tunnel was to be successfully completed. These were: (a) that the scope of the project should be clearly designed for once and for all. This was still a major irritation for TML – that the final specification had still not been established and so programming and planning remained incomplete; (b) Eurotunnel should face up to its commitments and pay what it owed; (c) relationships between TML and Eurotunnel must be improved. This was obviously easier said than done. There were now deep wounds on both sides and despite several personnel changes and attempts at reconciliation, the tension of the project and the immense amount of money at stake always forced the two sides apart. But the contractor said it wanted to sit down and make peace and seek out a deal without either side resorting to abusing the other. "Public denigration of companies is best left out of a contractual situation," said Costain; (d) funding of the project must be properly ensured by Eurotunnel. Eurotunnel's estimate of the final project cost stood at £8.05 billion, but this only included an allowance of £209 million for the fixed equipment for which TML had already claimed £810 million. Its available funds stood at £8.7 billion, but with costs rising dramatically TML was questioning whether more money might be needed to complete the project. With funds in place, TML argued, Eurotunnel might take a more reasonable attitude to solving the claims disputes.

"None of these conditions have yet been met," said Parayre. "It is not the 10 firms that are in trouble. It is the project."

Parayre went on to explain that TML believed it was owed £1.1 billion and that the contractors' cash flow would soon become negative. "We are not willing to fund this project," said Parayre.

Much time was taken explaining how fundamental changes to the work had had a knock-on effect on so many aspects of the fixed equipment. An increase in the width of the shuttles and a doubling of the electrical capacity were two examples of such changes which were so disruptive that they cost a small fortune. An increase in the tunnel diameter and a boost in the traffic estimates also affected work substantially.

TML estimated that it was now falling some £70 million a month short on the project, money it felt Eurotunnel should pay. "We are all determined to see Eurotunnel take on board the extra costs which are its responsibility. TML will not finance them," said Parayre. "Eurotunnel has destroyed the spirit of the contract," he continued. "This contract assumed a unity of purpose which would govern relations between the parties. Now they say the contractors underbid for the work. But at the time all the investments experts said our bid was generously costed. Can it be that they were all wrong, including 10 of the leading civil engineering companies in the UK and France?"

Warming to his task Parayre moved on to the tunnelling costs which, he said, had increased by 50 per cent. Half of this was due to poor ground conditions, the other half due to changes ordered by Eurotunnel, such as the increase in the tunnel diameter and the doubling of the number of cross-passages. Parayre said that TML had done well on the tunnelling in the circumstances. "Everyone recognised the merit of the companies in this difficult part of the site except our client who, in the guise of thanks, is trying to impose a fine of around FF1 billion."

This was indeed true. Eurotunnel had submitted a counter-claim against TML, on the tunnelling, for £100 million, alleging negligence by the contractor during the tunnelling work. Eurotunnel accused TML of "taking excessive risks, slackness with regard to inspection of the site and towards the preliminary studies" in a confidential report on the project. TML was shocked and offended. Neville Simms of Tarmac added that the contractor could not now predict when the contract would finish, leaving little doubt that the claims row and the finishing date were indivisible.

But for all the strong words and veiled threats, much of the mood of the meeting was conciliatory. TML believed it was time to talk and take the project on to its successful completion, and refused to threaten openly to walk off site or stop work. "We are running out of time," said Parayre. "We must settle our problems right now if we are going to save the project. Never since the beginning has the project been in such a serious situation," he concluded. And that was saying something.

17 History man

"I was going to have that weekend off to decorate the house."

Graham Fagg, TML tunnel miner

Graham Fagg thought he was in trouble. After a normal shift as charge-hand fitter for the UK service tunnel he had been summoned to the Transmanche Link management offices and that normally spelt bad news. His refusal that week to wear the garish orange TML overalls seemed to have reached the ears of the big bosses. "When they told me I had to report to the office I figured it was for disciplinary action," he said. "Just recently I had been in a picture in TML's in-house magazine, *The Link*" and I'd been the only one wearing denims rather than the orange uniform. I'm a bit of a rebel like that." But when Fagg got to the office he found he was not in trouble. Far from it. His name had been pulled out of a hat and he had won the right to complete the breakthrough between Britain and France and meet his French counterpart in an historic gesture in front of the TV cameras underground. Graham Fagg is typical of the breed of phlegmatic construction workers that are the backbone of the industry. He takes everything in his stride and is unlikely to get overly excited about pomp and ceremony. It's an attitude prevalent in civil engineering, borne of being prepared for anything and accepting that the job can so often disrupt plans and all you can do is get on with it in good humour.

TML had decided that the breakthrough honour would be granted to one of the tunnellers who had been on the project from the start of the tunnelling. In all, 70 names were in the hat and when Fagg's name was pulled he was summoned to be told the exciting news. Fagg was going to make history – how did he feel about that? "Well actually," he said, "I was going to have that weekend off to decorate the house and I was miffed that it was disrupting my plans. I wasn't due to work that weekend, and although I had been given one of just 30 tickets for the big breakthrough party at

Dover Castle, I had already given mine away because I wanted to get on at home." It is a fabulous reaction and one that could only come from the no-nonsense folk who make up the construction industry. When you don't get many weekends off, those that you have planned are special – historic tunnel breakthrough or no historic tunnel breakthrough. Despite the disruption to his domestic life Fagg agreed to take part in the ceremony – and for a while in December 1990 became the most famous construction worker in Britain.

Fagg is 46 and was born and bred in Dover. His first construction job was as a 17-year old when he worked as a general labourer on a relief sewer in Ramsgate, the first of many tunnelling jobs that he got involved in.

At first he was happy to go wherever there was work, but once he got married and the first of his three children, Sharon, arrived in 1968 he was keen to work close to home. He left construction to work in the Kent coal mines for three years, but was able to find a job on the 1974 Channel Tunnel attempt. Now qualified as a fitter and a miner he had skills which every tunnelling contractor needed. For 14 months he plied his trade on the tunnel attempt until the disappointment of the abandonment of the scheme and it was back to scouring the job market as near as possible to home. Not until his children had got older did Fagg finally seek work abroad and he joined the army of tunnellers building the new Cairo Wastewater system – a massive and very difficult tunnelling scheme.

When the Channel Tunnel was back on the cards in 1987 there was some disquiet that TML was poaching tunnellers from the Cairo contracts. Adverts appeared in English in the local papers calling for tunnellers to apply for jobs with TML. But Fagg did not see them. He did not have to. "In the tunnelling industry we all know each other and every scheme is well known. I knew TML wanted tunnellers and I wanted to be back home, so I applied." In August 1987 Fagg joined TML as a chargehand maintenance fitter in the service tunnel. The machine had not yet started its journey and had to be prepared ready to roll. Fagg was pleased to be back.

"It was really nice to be able to work on site and then go home at night rather than to digs. And with such a huge project we knew the work was going to last a couple of years or so which is great security for our sort of business. I was going to live a relatively normal life – although we were soon doing 12-hour shifts." In February 1988 Fagg was to get his first brush with the media when he was asked to explain to the then Prime Minister Margaret Thatcher the workings of the service tunnel boring machine. By then Fagg was a key member of the tunnelling team that was battling through the unexpected and demoralising wet ground. "It was a terrible time. Every day we worked soaked to the skin and covered in mud and the machine went practically nowhere. But you just get on with the work and keep going. Everyone lends a hand. There's no question of 'but that's not my job'. We all had to work closely as a team – you put personal feelings about each other behind you and just got on with it. Although the basic money is OK it's the bonus that is so good – and to get that we had to pull as a team." Fagg says the tunnellers just ignored all the rows going on at the management level. "We just used to laugh about it and get on with it. We never had any problems at our level." The blessed relief of the better ground soon came, as did the better tunnelling rates and finally Fagg's big moment at the breakthrough. "On the big day I was given a brand new orange outfit to wear and down I went into the tunnel to breakthrough. I remember doing the digging and Phillippe Cozette digging for the French from the other side. I thought that he was shifting a hell of a lot of muck – until I realised there were three of them working and just one of me."

The French had more room on their side, but in the small excavation hole on the UK side Fagg could hardly move with the TV crew packed in. "The whole thing was controlled by the adverts on French television which apparently could not wait. We had to stop working to time the breakthrough properly for French telly." The big breakthrough came and a beaming Fagg and Cozette posed for the photos and swapped flags and then Fagg climbed through onto the French side where French workers cut chunks off his new orange overalls as souvenirs. "I swapped hats with Cozette but then he gave me a piece of chalk on a ribbon as a present. I had told TML that I should take something with me to present to the French but they had said 'no, just climb through and shake hands.' Luckily, thinking on my feet, I handed over my tunnel ID tag." Fagg was then rushed away to France after a couple of glasses of champagne where he got the first immigration stamp on the French side. Then it was back on a plane to Dover Castle for the party. "I don't really drink much but the party was good. My wife and daughter had a fabulous time." Fagg had not taken any holiday on the project up to that point and figured it was as good a time for a break as any.

He left the project in June 1991 and has since worked in Newcastle and Lesotho before his present job on a sewer outfall in Norfolk. "I still can't find work locally which I would prefer. I applied to work for Eurotunnel doing anything but didn't get a job. So I'm up in Norfolk. It's not as if people remember me from the event – people in the tunnelling industry know me, but they always have. No-one else really remembers, although I'm about to go and do a programme about it all with the BBC." Despite his laid back attitude, Fagg now remembers the whole affair with affection. "At the time it was all in a day's work and I just got on with it. But looking back I realise now it was a very historic occasion." But he adds wryly: " I didn't get paid for that extra day though."

18 More talks but no trains

"1992 is the year that there just has to be agreement with Eurotunnel. We just can't go on."

Robin Biggam, chairman BICC

On the very day that TML tried to show its united face to the world's press in Paris, back in Britain one of those chiefs who had not made the trip to France decided to open his own lines of communication with the client.

Peter Drew, the chairman of Taylor Woodrow, wrote a personal letter to Sir Alastair Morton pleading for both sides of the dispute to "cool it". The news that Drew had acted independently of the other contractors was greeted with annoyance by some of the member companies. And it well illustrated a problem that TML was struggling to cope with. Obviously 10 powerful contractors would normally be expected to have different views on how to approach any particular problem, but TML as a consortium needed to find common ground in the dispute to maximise its negotiating strength. In most instances discussion and compromise would lead to the establishment of a party line to be toed by all. But on any particular issue someone is likely to be adamant that their way forward is the right way, and break ranks if that is not the accepted path of the others.

So while TML tried to convey a concerted message to Eurotunnel from Paris, the client's attention was clearly being distracted by a letter from London which, while not differing greatly from the main TML thrust, revealed a certain disarray in the ranks. And the client would no doubt have noted the conflict at TML exposed by the two different approaches.

Although this may not have seemed like a major rift, it was without doubt a sign of the struggle now going on within TML as the pressures on site mounted.

Of the 10 member companies some were in a stronger situation financially than others and therefore more willing to bluff Eurotunnel for as long as possible. Others were not so willing to drag out the dispute as they looked with concern at dwindling bank balances.

This pressure and the effect it had on the 10 companies naturally resulted in different ideas on how to react to developing crises.

So for the TML board to reach any sort of agreement on how to deal with Eurotunnel was remarkable. The trouble TML went to at the Paris press conference to show a united front was plain to see. Of the 10 men present, four were to speak and no more. The four sat in the middle of the table. The other six contractor representatives sat in alphabetical order by their companies names to ensure that no-one could read or assume a pecking order among them. And the line taken on all the many questions from the floor was a consistent one, rehearsed and agreed by men who rarely had the time to meet together because of their full-time responsibilities to their parent companies.

So it was a shock when Taylor Woodrow's Peter Drew acted independently and contacted Morton himself, even though his company colleague Tony Palmer was at the Paris meeting. And to pick the very day of the press conference to do so was surprising to everyone – TML, Eurotunnel and the media.

Drew's letter called for a "round table discussion" on the problems facing the project. But he added: "Of course it would need to be chaired by an entirely independent, positive thinking person of stature. Certainly the financial community would have renewed confidence in this great project and ensure that the necessary resources were made available to complete it if there were evidence of a new spirit of cooperation. I am sure that everyone involved in this great venture is anxious to find a way forward through what threatens to be a tedious and expensive impasse. What I am really suggesting, Alastair, is let's cool it and quietly work together for an equitable conclusion."

The letter was released to the press on the day after the Paris press conference. This also annoyed Eurotunnel, who had understood it would not be made public.

Morton's reply was that talks could start at any time through normal contract negotiation channels, but he gave no support to the idea of an independent outsider. "You do not need an escort to come in for a chat, Peter," he wrote. "Bring your fellow chairmen too. We have made it clear over the months we would prefer that to the hyper aggressive barrage we have received from TML."

Whether it was the Drew letter or the Paris press conference or just a coincidence is not clear, but talks started that October. Drew himself firmly believes his letter did help. "I think the conctractors had got a bit too tough...the press conference in Paris was all well and good but I could not see how it would help TML get its money."

Both Morton and most of TML's members accepted that Drew's idea to bring in an outsider to try to tackle the complex issues of the dispute would be impossible, and this was rejected. But the restarting of talks was very much seen as a step forward, although everyone played it down. After all, getting round a table, while representing some kind of progress, was hardly enough to resolve a dispute which both sides understood so well.

The fact that the talks went ahead meant that Eurotunnel's chiefs were actually willing to discuss TML's global claim for the fixed equipment, a breakthrough for the contractor. The position three months earlier had been grim, with Neerhout and the Project Implementation Division of Eurotunnel refusing even to consider the claim, and asking the disputes panel to rule whether the "global" approach was valid. Now they were talking about the claim. That was a start. Neither side was expecting a quick result but both said they would start with an open mind.

On a project where any progress was good news, both sides were again relatively buoyant approaching the end of 1991, although TML was to have one more setback.

In November 1991 Eurotunnel claimed a High Court victory in its battle to prevent TML stopping work on the cooling system, a weapon TML would have liked up its sleeve during the rows ahead. TML had asked the court to dismiss Eurotunnel's injunction application but Mr Justice Anthony Evans refused TML's request to suspend proceedings and awarded costs to Eurotunnel. TML was told that it must give 14 days notice of any stoppage of work, and although Mr Justice Evans did not think it necessary to award an injunction to Eurotunnel, he gave the developer leave to return to the court if notice of a stoppage was received. It was a small victory for Eurotunnel, and one that seemed irrelevant with talks apparently going well between the two sides.

But both sides were keen to maintain any semblance of advantage that they could over the other, and Eurotunnel's pleasure over the court victory was tempered by the knowledge that TML had decided to go to the Court of Appeal to try to have the decision overturned.

TML won the Court of Appeal decision in January 1992. The Appeal Court released TML from the High Court undertaking which it had been forced to give saying it would not suspend work in the project over this dispute. TML was awarded the costs by the three judges, Lords Justices Neill, Woolf and Staughton. This decision was made because Lord Justice Staughton said that an English court should not grant an injunction in a disagreement that all the parties had agreed ought to go to arbitration abroad. The construction contract allows disputes to go to the disputes panel and then to the International Chamber of Commerce in Brussels, so an application for an injunction – according to Lord Justice Staughton – had to be made to Brussels. But he did say that if the contract had only been subject to English law, then TML's claim to have the right to stop work because Eurotunnel had breached the contract would "have faced some difficulty". This time it was Eurotunnel's turn to appeal, which it did to the House of Lords.

At the beginning of 1992 the whole affair of the legality of TML walking off site seemed to be a bizarre and irrelevant sideshow. TML and Eurotunnel were engrossed in talks which were clearly making progress, and an air of real optimism was apparent.

Taylor Woodrow began the wave of positive reports on negotiations when it revealed that "talks were progressing at a faster rate and with a new air of realism". The statement was designed to calm market fears about the potential losses to contractors of the continuing dispute.

Behind the closed doors of the negotiating room the gap between the two sides was rapidly closing and the three key areas of dispute were all up for discussion – the cost of the project, the opening date, and the need for both sides to work closely together on the commissioning phase of the project to complete it in reasonable time.

Taylor Woodrow's optimism fuelled shareholders' hopes of an imminent deal, but TML and Eurotunnel both moved to dampen such high expectations, knowing just how much still needed to be agreed. "Some of the reports are a bit too euphoric," TML told *Construction News*. "But the talks are going well and there are certainly no grounds for despair." This was a far cry from the "deal imminent" stories being carried by some of the national press, but was an accurate assessment of the situation.

TML was willing to reduce its claim to meet an increased offer from Eurotunnel. Both sides had indeed shifted their position, but not yet far enough to meet in the middle.

Eurotunnel was keen to keep the amount of cash paid out to a minimum, but raised the possibility of offering the contractors further shares. TML, while preferring cash, was willing to discuss this as both sides had originally agreed to be open-minded and said that "everything is on the table". TML was also looking for a delay in the tunnel opening date, so that the penalties that it was due to pay for late completion would be waived.

For some time TML and other tunnel watchers – including the Maître d'Oeuvre and the banks' technical adviser – had maintained that the tunnel could not open on June 15 1993. Eurotunnel, for its part, was adamant that with the right atmosphere on site and total cooperation, that date could still be achieved. But as talks progressed some movement looked possible on the time front, particularly in light of the rolling stock delays which were set to hold up the commissioning phase of the project anyway.

While talks were continuing in such a positive light, TML and Eurotunnel agreed to postpone the contract disputes panel's consideration of TML's global fixed equipment claim, which had been referred to the panel in August by Eurotunnel as neither side could see any gain in a ruling while talks progressed.

In many negotiations the initial progress is often very good, giving rise to a perhaps false optimism that a deal is in sight. Tying up the loose ends and agreeing the last sticking points often proves much more difficult and time consuming. And indeed Eurotunnel and TML had found that at first, after so long not negotiating, they had enough common ground to make the early weeks of discussions particularly fruitful. But once they came to the nub of the problem, the detail of exactly how much TML was owed, talks slowed as the parties rapidly returned to their bunkers.

As the talks faltered, the old bad feeling between the two sides began to re-emerge. The smallest incident could affect the negotiations adversely. And to add to the tension another row was on the horizon.

On Monday February 9 1992 – with the talks slowing to a crawl - Eurotunnel announced that the opening of the Channel Tunnel had been delayed from June 1993 to "the end of summer 1993". Although the statement confirmed what everyone close to the project had suspected for quite some time, the stock market still did not like the news and Eurotunnel's London share price dropped 13p to 454p.

And TML did not like the way the news was presented.

Sir Alastair Morton would later fiercely deny that he had blamed the contractor for the delay, but the tone of the statement and of an interview he gave to Radio Four's Today programme made it clear where he thought the project had gone wrong. "The current rate of project progress, if maintained by TML, will not enable the Channel Tunnel to open for service as intended on June 15 1993," said Eurotunnel. "If the contractors meet programmes for installation and commissioning currently under discussion with them, the tunnel should be able to open at the end of summer 1993 with the launch of Eurotunnel's shuttle services."

"It is disappointing to confirm what has become widely accepted in recent weeks," added Morton.

TML believed the implication was clear and that it was being blamed for the delay. It was livid. One TML executive said: "I nearly crashed my car this morning when I heard Morton on the radio blaming us for the problems. It's outrageous. We have been warning that the tunnel cannot open in June for months."

A TML spokesman told the press: "Any implication that the delayed opening of the tunnels is attributable to construction difficulties on the part of TML is totally untrue."

TML said the reason for the delay was problems with the rolling stock deliveries caused by the changes to their design for safety reasons. "Even if we finished the tunnel three months ahead of schedule it would not make a jot of difference to the opening date, since we can't commission the system without an adequate supply of rolling stock and all three types of train are delayed," said the contractor.

The three types of train TML referred to all have a distinct role in the operation of the tunnel and were all being built by different manufacturers. They were also all behind schedule. The passenger shuttles, under construction by Bombardier/ANF/BN, were held up several months by design changes to fire doors. The HGV shuttles, being built by Breda/Fiat were delayed by up to a year by a continuing argument with the Channel Tunnel Safety Authority about whether the wagons should be open, semi-open or enclosed.

And British Rail and SNCF's through trains, which would run from London to Paris through the link, which were being built by GEC-Alsthom, were now not expected to run through the system until summer 1994 according to Eurotunnel. The big problem these delays presented Eurotunnel was the reduction in revenue caused by missing the lucrative summer 1993 market. This blow, when added to the ongoing delays on the building of a high-speed rail link from London to the tunnel and concessions given to the ferries on duty-free goods, was starting to put pressure on Eurotunnel's cash forecasts.

The banks were quick to show concern, but Eurotunnel maintained that although it had not yet calculated the full financial effect of the delay it was confident it had enough money to see it through. Morton strenuously denied that there were any plans or need for a further rights issue and certainly none before the tunnel opened.

The City knew the margin between the money Eurotunnel had available to it – even allowing for the extra £2.1 billion secured in 1990 – and the cost of the project was closing fast, and the banks were starting to demand some answers on the resolution of the £1 billion of claims. These still seemed to hold the key to the viability of the whole project.

All the pressure and the delay was affecting the financing of the scheme and Eurotunnel needed more than ever to cut costs and, if possible, reduce payments to the contractor. This was not conducive to fruitful negotiations. Eurotunnel's purse, if not ever exactly wide open, had now been snapped shut by the rolling stock announcement. There was not much cash left on offer for the contractor.

Not surprisingly, it was not long after the announcement of a delay in the opening of the tunnel that the noises from the negotiations were striking a less optimistic note. Both sides accepted that the balance of the relationship between TML, Eurotunnel and now, once again, the funding banks was very fine.

That delicate balance could easily be affected by a row over the responsibility for the delays.

So concerned was Eurotunnel not to upset the negotiations that Morton shocked tunnel watchers by suddenly – the day after his brash Radio Four interview – taking a more conciliatory line. He furiously denied that he had blamed the contractors for the delay and said that perhaps everyone was a little bit to blame for the project's problems. Despite this he had to admit that there were no developments in the talks with TML, "and that had to be bad news for everyone." The mood of pessimism that spring was acute because it followed a period when things had genuinely looked to be on the up.

The gloom was compounded when BICC – parent of TML member company Balfour Beatty – announced its 1991 results. It made a provision of £12 million against losses on the Channel Tunnel contract, the first British company to do so. But it was a clear sign that the other nine companies would follow suit.

BICC chairman Robin Biggam made it clear that the figure of £12 million only covered the period up to early 1992, leaving the door open for further provisions if necessary.

The French contractors had already begun to make similar moves, a more common occurrence in France.

The British five had been holding out until now, believing that enough of the claim could be settled to avoid the need for provisions and protect the price of the companies' shares. Biggam said he still thought the contractors would make a profit on the project, but said: "1992 is the year that there just has to be agreement with Eurotunnel. We just can't go on."

Whether the project could "go on" was now largely in the hands of the banks, who were becoming increasingly frustrated and worried.

Talks between Eurotunnel and the banks centred on the fact that the change in revenue forecasts and costs now meant that Eurotunnel was once again technically in breach of some of the conditions of the bank loans agreed. These conditions dictated that Eurotunnel could not draw down – or take for its use – portions of the money available through the loans if the margin between the funds available and the estimated project outturn cost became too narrow. This was now the case, as it was in the summer of 1990 during a cost crisis only resolved by Eurotunnel negotiating extra loans and a £500 million rights issue.

The banks needed to know the true claims situation as Eurotunnel saw it with regard to costs and revenues before they were willing to grant a waiver of their conditions to allow Eurotunnel to draw down a further £500 million due that May. In tense meetings Eurotunnel was able partially to calm the banks' fears and seemed to have convinced them not to withdraw their support for the project.

On March 18 1992 the four agent banks and their advisers agreed with Eurotunnel to recommend to the 220-strong banking syndicate that drawings from the loans should continue uninterrupted under an agreement to waive the loan conditions. As with the summer 1990 dispute, this was an essential recommendation. Without it Eurotunnel would run out of money, preventing payment to its contractor or settlement of any other outstanding bills. The project could only continue if funds could be drawn down, and on that March day there was every reason to believe there would be no difficulty. On March 26 all that changed.

The row over the blame for the delay in the opening of the project and the subsequent hardening of attitudes on the negotiations had led Eurotunnel and TML to ask the disputes panel to make its ruling on the global claim after all. This ruling had been deferred at the request of both sides during the brighter days when a deal seemed possible.

This ruling was announced on March 26 and it hit Eurotunnel, and the project, for six. The panel concluded that the fixed equipment element of the project should still be paid for on a lump sum basis, but that work affected by changes in scope or performance be valued at rates that were "reasonable and proper". If the two sides could not agree fair rates, then the panel was willing to fix the rates for elements of the work.

This idea of the panel fixing rates was new to Eurotunnel and one that it did not like. Rates had to be claimed in detail by TML and argued over by Eurotunnel, not just set by a disputes panel. Or so Eurotunnel thought.

If this part of the ruling was a blow to Eurotunnel then the panel's additional finding was a sensation. It ruled that Eurotunnel should triple its monthly payments to TML – from £25 million to £75 million – for the fixed equipment work, starting from April 30 and indefinitely after that. When the fixed equipment dispute was eventually resolved, the extra money paid now would be deducted from the sum owed, which the panel now believed was substantial.

Eurotunnel was astonished by the decision and immediately referred the matter to the International Chamber of Commerce in Brussels for arbitration.

Eurotunnel said the decision "raises issues of principle and findings in fact and law with which we disagree". Eurotunnel wanted the ICC to rule not only on the dispute but "the panel's surprising additional conclusion that it is entitled to stipulate the funding of work in progress." The client could not believe what it was hearing and was convinced that the panel had seriously exceeded its powers.

The decision – described by Eurotunnel in a hurried but shocked statement as containing "findings largely but not wholly favourable to TML" – threw the whole scheme into uncertainty and put the banks into retreat.

Despite agreeing to recommend a waiver on March 18 1992, the banks now decided in the light of the panel decision to rethink their position. What was now clear was that predicting a final cost for the project or even a reliable opening date was going to be extremely difficult. And certainly Eurotunnel's estimates of both looked rather conservative.

The banks retired to reconsider their options and withdrew their recommendation to the syndicate to grant the waiver which would have allowed the draw down of funds. Eurotunnel faced a cash crisis again.

TML greeted the ruling with a mixture of delight at Eurotunnel's discomfort but concern over some elements of the decision. The ruling was not as bright as it first seemed for the contractor. Although it looked to have guaranteed cash flow over the summer, Eurotunnel was already making it clear that it did not believe it would have to pay the extra £50 million a month while the matter was referred to arbitration, which, as Eurotunnel pointed out, could take a long time.

TML argued that the contract did not allow a party to set aside the panel's ruling while it went to arbitration, and another issue for the lawyers to dispute was created.

TML's case for submitting a global claim for the work had also taken a knock, with the panel maintaining that the lump sum system should stay in place. So, despite the panel being willing to fix elements of the work, item by item justification of the dispute was still very much on the cards and this was a scenario TML was still very keen to avoid.

The first extra £50 million payment was due at the end of April, but on April 23 the Eurotunnel board authorised the company to resort to arbitration and so hold back on the extra payment. The next day Eurotunnel presented its latest progress report to shareholders in France and the UK, and it did not make happy reading.

Despite assurances that the tunnel could still open in September 1993, it contained the first warnings that the programme could slip into 1994. John Neerhout, project chief executive, said: "If everyone involved got together satisfactorily we could make that date. Our concern is that if we don't get together right away, the date could slide."

The report admitted that the panel ruling had put pressure on Eurotunnel's relationship with TML and that if the two sides could not get together and cooperate the commissioning phase of the project would be severely hindered, pushing opening further and further back.

Morton did say that the claims and costs of the French terminal works had been agreed to everyone's satisfaction. But there were still outstanding claims over the UK terminal and the tunnelling.

Morton's victim for personal criticism at this annual report was GEC's managing director Lord Weinstock – a pleasant change for the contractor and the Intergovernmental Commission.

GEC-Alsthom was responsible for the supply of the £600 million fleet of Transmanche Super Trains to British Rail and SNCF. But, Morton said, there were problems. "I'm afraid Lord Weinstock and his French colleagues in GEC-Alsthom are going to be late with British Rail and SNCF's trains. They won't get running until the early part of 1994."

Despite the waves of bad news, the remarkable Sir Alastair managed to remain bullish throughout the presentation of the report to the press, concentrating on those areas Eurotunnel could fight and do something about and giving short shrift to any journalist daring to ask searching questions about money and delays.

Morton's coup de grâce was the announcement that Eurotunnel was preparing to sue the British and French governments for breach of contract, based on the signing of the 1987 concession document.

Eurotunnel claimed up to £200 million because the two governments had allegedly failed to honour agreements to build adequate rail infrastructure at either end of the tunnel, and that they had increased the cost of the project by imposing excessive safety restrictions. It was a claim that looked incredible but Eurotunnel was serious. And it persisted until it received some satisfaction when, in December 1993, it agreed to drop the claim on the basis that its concession to run the tunnel and hence earn revenue from the project was increased from 55 to 65 years.

But despite Morton's public shows of bravado, behind the scenes the problems with the banks were mounting. No waiver had been granted and Eurotunnel was rapidly running out of money. The banks had a stranglehold on Eurotunnel with the date for the next drawdown of funds the end of May. The four leading banks – National Westminster and Midland Bank of Britain and French banks Crédit Lyonnais and Banque Nationale de Paris – had a lot of power that month. And they used it immediately to order Eurotunnel to pay £50 million to TML on April 30, equivalent to the sum ordered by the panel ruling which Eurotunnel had been saying it would not pay. Eurotunnel maintained that it still did not accept the panel ruling but had no choice, with the waiver not awarded, but to do as it was told by the banks.

The payment prevented further legal action by TML which was threatening to go to court if Eurotunnel refused to pay the money.

Eurotunnel reserved the right not to pay the next installment and this prompted TML to draw up contingency plans if the money was not forthcoming. Suddenly the Court of Appeal decision giving TML the right to strike came back into play, as TML lined up areas of work that it would slow or stop if it was not paid the extra £50 million each month.

In a letter to shareholders that month Eurotunnel left the door open to make further payments "when necessary", as its hard line evaporated in the face of the pressure from the banks. What the banks wanted was extra security for their loans, and were pushing for Eurotunnel to agree to another rights issue some time in the future, ideally after the tunnel had opened. Unfortunately that opening date was still not clear. In fact, it seemed less clear now than it ever had been.

The other vital factor for the banks was that they needed certainty about the cost of the project, and so they pushed Eurotunnel to make concessions to get a deal with TML over the claims.

At the end of May 1992 Eurotunnel made TML an offer of £980 million cash in settlement of the fixed equipment work, which was originally to cost £620 million but for which TML had claimed £1.27 billion plus a

£160 million management fee (all these figures are in 1985 prices for ease of comparison). This deal was rejected by TML, which then – after trying to shift Eurotunnel further – asked the disputes panel, as it had promised in its March ruling, to set the price for the fixed equipment claim, which had now been the subject of dispute for more than two years. Eurotunnel, for its part, asked the panel not to make such a ruling because, it said, the International Chamber of Commerce had been asked to rule whether it had the power to fix costs on the project. Eurotunnel also referred the situation again to Brussels.

The banks were no nearer their answer on project costs, and spent considerable time in hard bargaining with Eurotunnel. In those talks Eurotunnel was able to convince the banks that if it was under too much pressure from the banks to make a deal, the project's costs would rise because the final agreement would favour TML.

Grudgingly the banks accepted this and agreed to give Eurotunnel breathing space as long as it continued vigorously to pursue a settlement with the contractor. So despite the breakdown of the talks the banks granted Eurotunnel its waiver right at the end of May. The deal advanced the project a further £500 million to save Eurotunnel from insolvency.

As part of the deal Eurotunnel also paid its second £50 million extra payment to TML at the end of May – a payment it could now afford more easily than in April.

The banks insisted that the waiver be renewed in August 1992, but in the meantime they wanted Eurotunnel and TML to negotiate a deal once and for all.

The negotiations in that summer of 1992 were remarkable for the fact that once again Sir Alastair Morton was face to face with the top executives of the 10 companies.

Despite the introduction to the project of TML's chief executive Jack Lemley in 1989 and Eurotunnel's project chief executive John Neerhout in 1990, no deal had been struck between the two men.

By 1992 it was time for some new faces to try and find a breakthrough – and Morton stepped in for Eurotunnel while TML fielded chief executives from the 10 contractors.

When Neerhout was brought into the scheme one of his functions was to act as a buffer between Morton and the contractor, and some of the member companies wanted him out of the project altogether.

So it was somewhat remarkable to find him back, less than three years later, negotiating face to face with some of the men who had wanted his head on a plate. Progress was made. Deals were struck on the UK terminal and the target costs work which covered the tunnelling.

The UK terminal, according to TML, had changed fundamentally due to variations in the design and a rethink on the way the terminal would operate and how much traffic it would take. Despite this TML was able to itemise its claim in some detail and was paid in line with the amount it would have expected of a client on a part of the job which was, after all, standard civil engineering.

The bone of contention on the tunnelling was the problems with the ground conditions and changes in the tunnel diameter. A compromise was agreed on this too. But these deals still left the big claim – the one for the fixed equipment – outstanding.

Eurotunnel had a certain amount of cash available but was keen to push the idea of topping up the deal with shares or "convertible paper". Several instruments in the form of shares, stocks and certificates were on offer to TML, each tying the contractor into the success of the project. The more successful the project, the more its paper would be worth when it was ready for cashing in.

It was a complex area of negotiation, and a sensitive one. Diluting Eurotunnel's equity by issuing shares to TML might not have been popular among shareholders, and any deal was conditional on their approval at an extraordinary general meeting. Eurotunnel was not best pleased, therefore, when out of the blue TML released a statement to the press in June 1992 saying: "Eurotunnel has proposed to pay TML part of its debts in Eurotunnel shares or other, similar, securities. To help Eurotunnel, the companies are ready to consider taking payment of a limited part of the money they are owed in this manner."

The reason for the statement and its timing is unclear. TML may have wanted to let its member shareholders and the public know of its willingness to compromise in the deal, but in doing so had revealed elements of the negotiations which Eurotunnel would rather have kept to itself at this stage.

Morton sent a letter to TML chairman Pierre Parisot, accusing him of jumping the gun and calling the statement "less than proper".

He did, however, admit that negotiations included discussions of a share issue, but stressed that Eurotunnel would do nothing without the agreement of its shareholders. It was another annoying episode for the negotiators, breaking the uneasy truce between both sides and stretching the patience of all those trying to do a deal.

But despite any damage caused by TML's statement, the summer talks progressed and in fact came close, but not quite close enough, to settling the whole affair.

In August 1992 the two sides seemed to have agreed to a non-binding "Heads of Agreement" document which would be the basis for negotiating a binding settlement. This deal would include a revised contractual programme to completion in December 1993, a further three months later than expected but in line with TML's site progress. Also in the agreement was a protocol to establish the cooperation required between the two sides during the crucial commissioning phase, plus the tests required to achieve completion on a phased opening basis. Everything seemed to be in place to establish once and for all peace on the project and get the tunnel running by the end of 1993.

Eurotunnel had increased its £980 million cash offer slightly and was adding in various combinations of shares and redeemable stocks, subject to the approval of the shareholders. This represented a deal of around £1.1 billion and, while not yet agreed by TML, seemed reasonable enough. At a dramatic meeting on August 13 1992, Eurotunnel initialled the Heads of Agreement and waited for TML to do likewise. But in a late change of heart TML got up and walked out of the talks. TML's problem with making a big decision was exposed.

Although the negotiators seemed happy with the deal, others were not. The more powerful French contractors, not so troubled by dwindling cash reserves, wanted to brave it out and go for more money, money they felt they were owed. So, at the last minute, they refused to sign.

Negotiations on the fixed equipment claim would have to reopen, and Eurotunnel again had to face its banks with the news that there was no deal, no definite opening date and no agreement of the final project cost.

The agent banks and the European Investment Bank met TML and Eurotunnel on August 24 and told the meeting that Eurotunnel's offer was a generous one and should not be increased unless there was a full technical audit of the project – something which TML did not want. This was a key meeting for Eurotunnel. With the backing of its banks it had the strength to stand up to the contractor and fight. Earlier in the year, with the panel ruling going against the client, the banks saw Eurotunnel as the agitator and bullied the company into making concessions. Now the banks had seen Eurotunnel strive hard to make a deal. They had watched closely as Eurotunnel juggled its available assets to present TML with a generous package, and they had watched in disappointment as the deal had been thrown back in the faces of the Eurotunnel negotiators. The banks were starting to change sides. Yet again the balance of power on the project was about to shift.

19 Brussels backs Eurotunnel

"There has been some improvement in progress recently – and about bloody time too."

Sir Alastair Morton

All eyes at the Channel Tunnel – this great British and French project – were now on Belgium. With the dispute over project costs still raging and the talks at a standstill, the next chance of moving towards a deal rested with the International Chamber of Commerce in Brussels.

The ICC was in the spotlight because Eurotunnel had referred the contracts dispute panel's March 1992 ruling to arbitration, as it was allowed to under the contract. Eurotunnel was asking the ICC to rule on three points: whether the panel was right to order Eurotunnel to pay £50 million a month extra to TML; whether TML could ask the panel to fix the price of elements of the fixed equipment; and whether the panel was empowered to set those prices if asked to by the contractor.

The arbitrators received submissions from both sides almost immediately, heard final pleadings from both sides on September 3 and 4, and agreed to rule in October 1992. Each side desperately needed a good result to boost its position in the negotiations.

TML was confident that if the panel was allowed to fix the costs, then the contractor would receive a good settlement for the fixed equipment without having to submit an impossibly detailed claim.

TML believed that that deal would be better than the one put on the table by Eurotunnel in August, and was worth pursuing before settling anything between the sides. For its part, what Eurotunnel wanted was the £50 million a month ruling lifted. It also needed more than just a deal on cash for the fixed equipment. It wanted an all-embracing "peace" document, which would cover the fixed equipment, a finishing date for the project and a schedule for cooperation on testing and commissioning.

This last point was crucial. The commissioning phase had been thrown into disarray by the non-arrival of the trains and Eurotunnel's announcement of a phased build-up of services from the opening date and on through 1994. TML held some strong cards here. Nothing in their contract called for phased opening of the tunnel, and the contractor still believed that, disappointing as a late opening would be for its member companies, it could be devastating for Eurotunnel because of the financial pressure from the banks.

But this was a dangerous game to play, and a split was emerging between the contractors over when and whether to accept a deal.

The doves, who wanted to accept Eurotunnel's August offering, were mainly the British contractors. The French were the hawks. They were convinced they could call Eurotunnel's bluff and make life so difficult that under pressure from the banks it would raise its offer, perhaps by as much as £150 million.

In September 1992, as they all awaited the ICC decision, Eurotunnel was embroiled in long and detailed talks with the banks. The waiver which allowed Eurotunnel to draw down £500 million in May had run out at the end of August, when a deal was meant to have been struck between the two sides.

Failure to do this was once again jeopardising the future of the project, and Eurotunnel again had to convince the banks to continue to waive the loan conditions to allow work to continue. But Eurotunnel had to do more than convince the four agent banks to back the scheme.

They also had to give these four the arguments to take to the 220-strong banking syndicate who, from afar, could only see desperation, disarray and delay. The syndicate wanted a deal between Eurotunnel and TML. But Eurotunnel's argument to the four leading banks was that TML knew Eurotunnel was being pressed to conclude a deal and was therefore taking a hard line.

The four agent banks were sympathetic to this argument, having been privy to the negotiations for several months. In their opinion Eurotunnel's last offer had been generous. But there was more work to be done to convince the other banks that Eurotunnel was still a good investment.

One of the factors that helped Eurotunnel at this stage was that the banks had few options. They had invested £6.8 billion in the scheme already, an amount of money they were hardly likely to write off.

On site the contractor was within a short distance of completing the work on a truly great civil engineering project – to stop now and let the tunnel fill with dripping water and rats was unthinkable.

To sack Eurotunnel did not make sense either. That would mean leaving the contractor to manage itself to completion and the banks to take over the project directly which also did not make any financial sense.

Eurotunnel was gearing up for the opening and was almost ready for business to start.

So the banks had to press ahead. But not without making life a bit difficult for Eurotunnel who spent September updating its financial cost and revenue forecasts, taking into account the phased opening.

While some of this information could be accurately assessed, without an agreement with TML or an opening date it was difficult to estimate the total cost to completion, the figure the banks really wanted.

By now Eurotunnel was starting to build up its operating staff, all of whom added to its outgoings. The further the opening was delayed, the more these costs counted against the cost to completion.

Eurotunnel's cajoling and persuasion again paid off, however, when, on September 30 1992, the agent banks notified the developer that the necessary majority of the syndicate had consented to a further waiver. But they had only granted enough funds to take the project to November 30. This short waiver was given to allow time for the syndicate to receive and study plans for the further financing of Eurotunnel. It was becoming increasingly clear that another rights issue was on the cards – and detailed figures were required by the banks before they would grant Eurotunnel further, and maybe longer-term, waivers to Eurotunnel.

While the banks and Eurotunnel battled on, and TML and Eurotunnel continued to talk, progress on site was still sluggish. The problem was a conflict of interest between the installation teams, which were doing the final fitting of sections of equipment, and the commissioning teams which were beginning to get items of the mechanical and electrical services up and running. Some areas of the tunnel had already been handed over to the commissioning teams even though they still required further installation work to be carried out. The installation teams and commissioning teams, to put it bluntly, started getting in each other's way, hampering progress and causing rows on site. The situation was becoming so awkward, with an unhealthy rivalry developing between the two different teams, that TML moved to change its management in an attempt to improve the on-site communications between the two sides.

Site commissioning director Tom Clay was given the task of smoothing the operations, and both teams were ordered to report to him. He would then make all decisions on conflict of programme, prioritising the work to maximise the speed of installation and commissioning. He in turn would report to commissioning director Marc Chatelard on commissioning and to operations director Peter Allwood on installation.

Tom Clay had the physical advantage of working on site and hence was able to make snap decisions. Previously the decision makers were based in the contractor's central liaison office and were distanced from the problems and priorities. But someone did not like the restructuring. An anonymous note sent to newspapers contained a stinging rebuke of TML for making the changes. The note was headed "Wholesale changes at Transmanche Link" and read: "Following intense internal power struggles, there is now a new organisation structure at Channel Tunnel contractors, TML. Out go the complete technical group including the commissioning and commercial teams, with the two construction groups taking over the central liaison role. Even though the French and UK construction groups rarely see eye to eye.

"This change means that the highly complex and vital commissioning phase of the project will be handled by civil engineers, not the specialist project teams that developed the sophisticated systems."

The note went on to explain that the removal of the commercial team also meant that "non-specialists...concerned only with civils" would be responsible for the accounts of the subcontractors, "including many involving subsidiaries of the 10 parent companies". The note signed off: "At a time of delicate negotiations with client Eurotunnel these are certainly daring moves."

A TML spokesman dismissed the letter as "absolute rubbish" and said: "This letter is completely untrue. There have been no intense power struggles, just a change in the management links in the team. I can only imagine that whoever wrote this did so before knowing fully what was planned."

It was another small but strange chapter in the project's history. Despite the denials, there was clearly tension on site between the commissioning and installation teams and, if the note was true, between the construction team and its commercial and commissioning arms.

It was little surprise that progress on site was poor. And until TML's morale could be boosted, then progress was unlikely to speed up.

Just two weeks after TML's management changes the ICC made its long-awaited ruling on the dispute – and dealt a further blow to TML's flagging spirits. On the subject of the £50 million extra interim funding which was TML's financial lifeline, the arbitrators found that the disputes panel decision was "wrong in law, unjustified by the evidence and, therefore, by the letter and spirit of the contract."

Eurotunnel had made four extra payments to TML since the panel decision in late March, totalling £200 million. But the arbitrators' decision meant those payments would stop. The £200 million was to be held on account against future payments made in settlement of the fixed equipment claim.

The Brussels chamber also ruled on whether TML could appeal to the panel for costs to be fixed. On this point the arbitrators took the view that they were not willing to stop TML going to the panel, even if they had the power, which was not clear. But they issued a warning that any panel proceeding should be careful not to rule on any issues still being considered by the arbitrators, and should not make decisions that might have to be overturned by the arbitrators in future. In the meantime the ICC – having been asked to deal with these matters as a matter of urgency – decided it was not going to rule on the principles of the panel fixing the prices of the fixed equipment until the New Year. So although TML could go ahead with its appeal to the panel to fix costs, the ICC was effectively tying the panel's hands by saying that any ruling it might make must not be such that it could be overturned later – a strange state of affairs.

The arbitrators' ruling reached Eurotunnel's ears almost at the same time that the company was informed that the banks would continue the waiver for two more months until the end of November.

The double dose of good news was particularly timely for Eurotunnel, as it arrived on the eve of one of its twice-yearly "state of the nation" addresses which accompanied the announcement of the company's yearly and half yearly results.

The April 1992 meeting had been a nightmare for Eurotunnel, coming days after the panel ruling which made the company pay the £50 million extra a month. The October meeting was entirely different, with the banks backing the project and the panel overruled. Veterans of Eurotunnel press conferences immediately noticed something different about Sir Alastair Morton at this London presentation. While never putting aside for long his pugnacious and critical style, there was a more relaxed air about the man. This was almost unnerving to journalists, braced for the normal 10 rounds of jousting. He even had a warm smile for the assembled throng. Despite much worrying news on the scheme, Morton looked like a man who believed he had at last gained the upper hand in a long fight.

His report showed that he did not expect the tunnel to open at all until December 1993 and that a full fleet of trains would not be running until summer 1994. And, he said, even those dates remained uncertain.

It was the contractor's lack of progress that had caused the problems, he continued. "We did not think it possible that they could make such a hash of this work (the fixed equipment)," he said. "You would have thought after five years they would have been ready for the services installation. But they weren't.

There has been some improvement in progress recently – and about bloody time too."

He explained that although Eurotunnel was striving to sew up a deal with TML, he was willing to continue to resist claims that TML could not justify. There was no deal yet, and the gap now stood at around £130 million.

Morton told the Eurotunnel shareholders that shares and securities were being offered as part of the deal, but he could not give any details as no deal had been done. He also let it be known that it looked likely on current projections that there would be a third rights issue in 1994 and 1995.

With the business part over, Morton left the bombshell for the post-match press conference. In an interview with *Construction News* he said that the deal that Eurotunnel had left on the table in August was its final offer, and added: "unless TML agrees to a deal in the next few weeks there won't be a deal at all."

If they did not accept, Morton said, the matter would go to long and costly arbitration, and nothing would be paid for years. "Our banks believe this offer (£980 million in cash plus another £200 million in non-cash payments such as shares and stocks) is very generous, in fact too generous," he said. "In the next few weeks we must prepare our submission to the banks on how the project may proceed – either the peaceful agreement route or the argument route where each bit of the project has to be ground out slowly."

The key to the shift in Morton's attitude was that the banks were no longer demanding an agreement with TML as a condition of continuing to fund the project. Exasperated by the delays, the banks were willing for the first time to look at Eurotunnel's projections of the level of costs and revenues even without a deal with TML, but assuming a long battle after the project was completed to sort out the finances.

Morton now no longer saw a deal with TML as a priority because of the banks' vote of confidence. And with the banks behind the scheme and the arbitrators ruling in its favour, Morton described the situation as "six love, six love, six love" to Eurotunnel. This was an exaggeration – the match was clearly not over although TML may have been a set down.

Without a deal the opening date was uncertain. Eurotunnel was talking in terms of a phased opening either side of Christmas 1993.

But the commissioning would still require close cooperation even for that deadline. Any slowdown on site could push completion well into 1994, and that could still damage Eurotunnel however confident Morton might appear.

Following Morton's ultimatum TML held top-level meetings in Paris and London to decide its next course of action. Eight of the contractors at that point were said to be keen to settle. With pressure building up in the domestic market, tumbling profits and the lifeline of the extra £50 million stopped by the arbitrators, there seemed no other choice. But two of the French contractors were having none of it. They still believed they could squeeze more money, which they felt they deserved, out of Eurotunnel. The ruling on the global claim was still to come. And they could use the ultimate weapon of slowing down work, so stepping up the pressure on Eurotunnel again.

But TML's decisions needed to be unanimous. Without the backing of all 10 contractors, no deal could be struck with Eurotunnel. So there was no deal.

Eurotunnel, while continuing to talk to TML to try to find common ground, was more concerned with sorting out arrangements with its bankers. Rising costs and lost revenue due to the delayed opening had forced the outturn cost of the project up, to around £10 billion, and Eurotunnel still had to propose to the banks how it intended to raise such a sum and particularly how it intended to approach funding after opening.

This plan, however, was naturally affected by the lack of an opening date. Eurotunnel told the banks that it still believed a service could be running in December 1993, but only with the full cooperation of the contractor. But for the sake of the funding requirements, Eurotunnel made its calculations assuming no deal with TML.

The dates used were those estimated by the banks. These put the start of freight shuttles and railway freight services at February 1994 and passenger shuttle services at April 1994.

No scheduled through passenger train service would be operated by the railways until the third quarter of 1994.

Using these estimates Eurotunnel presented the banks with its new financial projections in October and November 1992. By then Eurotunnel's August offer to TML had formally lapsed, as Morton had threatened it would, on October 20. Despite this, informal discussions still continued on both sides through November, focusing on the need to complete as early as possible. The main difference in the talks now was that some of the pressure had been lifted from Eurotunnel. Once TML members had agreed – or disagreed – among themselves in Paris not to accept the offer on the table, Eurotunnel knew it had to arrange its financial affairs without an end to the row.

Sure enough, at the end of November 1992 the banking syndicate agreed to grant Eurotunnel an 18-month waiver until May 1994, a huge boost. The company described this as "a tremendous mandate from the banks, telling us to go ahead and get the project finished."

There were, of course, conditions attached, the main one being that Eurotunnel had to submit to the banks more detailed funding proposals towards the end of 1993 in order to gain formal approval.

But this did not concern Eurotunnel excessively. The firm said the waiver justified its belief that it would complete the Channel Tunnel, and fears that the company would fail in the process were receding fast.

Many believed that the granting of the waiver would herald a settlement of the claims with TML. The thinking was that the banks would only have given their support if they were fairly sure a peace deal was close to completion.

With the waiver granted and the financial pressure lifted from the client, funds would be a bit more relaxed and the key documents signed, the argument went. Not a bit of it. The banks had been convinced that Eurotunnel had offered enough in August and were horrified by TML's rejection of that deal.

Now as they looked at the arguments again with Eurotunnel, they felt the August offer may have been too generous. TML had missed its chance of the £980 million cash plus equity when the deal lapsed in October.

With the banks' backing, Eurotunnel reassessed TML's position and decided it could not see the contractor justifying more than £900 million plus paper. Eurotunnel duly reduced its offer to £900 million in cash plus the paper element which totalled another £270 million according to Eurotunnel. This offer was intended not only to pay for the fixed equipment but also to provide a comprehensive deal on all issues relating to the completion of the project.

On December 14 the TML member companies issued a statement that the talks had once again come to a standstill. It said: "The member companies of Transmanche Link announce with regret that they have been unable to reach an agreement with their client Eurotunnel on the payment to which they are entitled. The member companies today confirm their unanimous view that the total sum proposed by Eurotunnel and the structure of the proposals, particularly the quantity and nature of the non-cash financial instruments, remain unacceptable. Negotiations are therefore at a standstill. Recognising Eurotunnel's financial constraints, we have gone a considerable way towards accommodating ET's proposals – to the point where we have come within a short distance of reaching an equitable agreement. However, not only is there a gap remaining to be bridged but the uncertainties contained within the non-cash financial instruments make us unable to accept this proposal on behalf of our shareholders."

But the statement went on to say that an early settlement would have produced the "strong and cooperative effort which is needed to complete the project with all due speed." This was immediately interpreted by Eurotunnel as being a veiled threat by TML to slow the work down as there was no deal.

Morton replied "The way is still open for constructive negotiations. But we note with particular regret the indication at the end of the statement that they may withhold cooperation, thus delaying completion. Our first priority remains the completion of the project as early as possible, as jointly agreed with TML, so that it can open for revenue service in December 1993."

But sources close to the project now doubted that any opening could take place in 1993. One senior project adviser said; "We have just started the commissioning phase of the project when we must work so closely together. No deal means that this will be hindered and so the programme is in danger. Eurotunnel's offer was aimed at not just paying for the work but also buying goodwill and cooperation between the sides, so that trains would run through the tunnel at the end of the year. Now that has been lost."

Despite Eurotunnel's stronger position following the backing of the banks, the waiver had been granted "subject to there being no serious material change in the project." With TML now unofficially making it clear that they could see no opening until summer 1994, and that a phased opening would not happen without their say-so, that serious material change was developing.

A year that had started with a Taylor Woodrow statement that a deal was imminent had finished with the two sides further apart than ever before, and no opening date for the tunnel.

20 Bombardier's bombshell

"He asked us to build the contract. We have done it. Trains will be running through and people will be waving. All we want to do is get paid."

Colin Parsons, chairman, Taylor Woodrow

The timing of the House of Lords ruling on whether contractor TML could walk off site or not was extremely unfortunate, coming as it did in January 1993 with formal talks between the sides at a standstill and Eurotunnel warning its contractor about slowing the work.

The case had been running since October 1991, when TML had threatened to stop working on the project's cooling system. TML argued that it had not been paid for the extra work demanded by Eurotunnel when the client upgraded the system's specification.

Eurotunnel had gone to the High Court and had obtained an injunction to stop TML trying such a move. At the Court of Appeal, TML had succeeded in getting this ruling overturned, but Eurotunnel had taken the issue to the Lords.

In fact, work on the cooling system was long since finished, rendering the original dispute somewhat irrelevant. But the principle was still there to be fought over fiercely. So when the House of Lords ruled in January 1993 that Eurotunnel could not prevent TML walking off site, it was still significant.

Eurotunnel played down the importance of the ruling, claiming it was now a legal instrument being debated by lawyers. But TML saw the decision as important, particularly if it was going to be accused of slowing work or even stopping work on site. As one site source said: "Eurotunnel has lost a weapon in the battle. It is not clear what effect this will have on the negotiations, but it is clear that Eurotunnel is upset at losing the case because it wanted to forget the threat of TML stopping work. It is in neither side's interest to string out the project – but Eurotunnel is particularly keen to get the project finished."

Buoyed by the decision, but still frustrated by the deadlock, TML held a series of meetings to decide what to do now to pressurise Eurotunnel if talks did resume.

TML was again shuffling its management team and both contractor and client hoped that new faces might precipitate a breakthrough. Dumez's Jean-Paul Parayre, chairman of the French arm of TML and therefore co-chairman of the consortium, stepped down at the end of January to be replaced by Philippe Montagner from the giant French contractor Bouygues. As he left, Parayre boosted hopes by telling French newspaper *Les Echos* that the basis for an agreement had been forged and a deal was now possible. He said: "The foundations of an agreement exist, although there is still a gap concerning cost. The gap is not of such a size, considering the project, that it could justify a failure."

The change at the top gave new impetus to a round of informal talks, in which Eurotunnel's offer reached around £935 million plus "paper" worth around £270 million. Once again these talks came close to an agreement. This time the stumbling block was TML's wish to cash in its new shares quickly to raise money.

TML's holding under the deal would have been no more than 15 per cent, and probably nearer 10 per cent, of Eurotunnel's shares. But the client was concerned that the contractor would sell those quickly, flooding the market and so reducing the price prior to the rights issue it was now planning for 1994. Eurotunnel therefore wanted conditions imposed on the paper element of the deal dictating when the stocks and shares could be cashed in, and this was not popular with TML's member contractors.

Talks through February 1993 ground to a halt at the end of the month when TML, according to Eurotunnel, "went into reverse", with the non-cash element of the deal the main stumbling block.

Again it was the turn of BICC – parent company of TML member contractor Balfour Beatty – to start the construction reporting season with its 1992 results and break the bad news to its shareholders. "There are no serious discussions taking place (on the Channel Tunnel deal)," announced BICC chairman Robin Biggam.

BICC had made no interim provisions in October because, like everyone else involved in the scheme, the company believed a deal was imminent. Balfour Beatty at that time was backing a deal along with most – but, crucially, not all – of the 10 contractors.

But the inability of TML to get a unanimous vote among the 10, plus the reduction in the terms being offered, had scuppered any hopes.

"Here we are six months later," said Biggam, "and there are still no signs of settlement." The company added £8 million more of provisions against the scheme to last year's £12 million. Biggam said that TML would go to arbitration if necessary, but that this would take years. He drew attention to the vital commissioning phase of work which was about to start on site saying, that a smooth commissioning period was vital but was only possible if a deal was struck. "There is going to come a time, and we are rapidly approaching it, when there has to be complete cooperation at the commissioning period," he said. "This is the most sophisticated piece of rail track in the world."

BICC's decision to make provisions against the tunnel was a surprise to the City and shareholders. But it reflected the lack of confidence that the 10 companies had that a deal on the £1 billion claim was close, and the sure knowledge that the project would only keep going if TML soon started funding the work. Last time this looked imminent the panel award of £50 million a month eased the pressure.

Now, with the interim award scrapped and Eurotunnel's monthly payments once again not covering the work done on site, the point was going to come when TML's reserves would be used up and the contractor would have to start paying for the work itself.

TML had already stepped up work to try to keep installation up to speed. But the commissioning process was in disarray. Commissioning had to be carried out to the satisfaction of the Channel Tunnel Safety Authority and, of course, Eurotunnel.

The original plan was for TML to install the system, commission it, and then carry out commissioning tests with Eurotunnel and the safety authority present to approve the work. But to make up lost time the new plan was to try to overlap some of this commissioning work to prevent it being carried out twice, once by TML on its own and again with Eurotunnel and the safety authority. This of course would rely on close cooperation between the management of TML and Eurotunnel to ensure that the right people were in the right place to check and sign off various sections of the system.

A further complication was Eurotunnel's wish for a phased opening. This significantly changed the way commissioning was carried out because it had to tie in with when trains and wagons were available on site. This in turn depended on progress at the manufacturers, which was delayed due to design changes. But to alter the commissioning phase so dramatically required TML's goodwill, something which was in short supply as arguments raged off site.

So knowing the risks ahead for the project if – as seemed likely – there was no deal, Biggam and BICC made the extra provisions and threw in a plea for peace for good measure. Without an agreed commissioning procedure, opening could yet be years away.

Eurotunnel knew this and so did TML. And it was still the contractor's strongest negotiating point. What TML had to decide was how to use the House of Lords ruling and its ability to hurt Eurotunnel by slowing work to good effect.

Once again there was a split between the hawks and doves among

TML's 10 members. But this time the doves were given one last chance to go to the negotiating table.

TML told Eurotunnel the consequences of not signing an agreement with the commissioning phase now starting on site, and put it to Eurotunnel that to increase its offer and reduce the conditions on disposal of the shares would be worthwhile in the long term.

TML proposed a management structure which, it believed, could see the work finished and commissioning carried out in time for a December 1993 opening. This was a variation on TML's approach up to that point. Previously the contractor had been looking for a traditional settlement of a traditional claim for extra work, while Eurotunnel sought a more ambitious peace deal.

TML now responded for the first time with its own peace deal, but still wanted to be paid a higher price than Eurotunnel was prepared to offer.

Although TML muttered darkly about months of delay if its new deal was not accepted, Eurotunnel balked at what it saw as blackmail, dug in its heels and refused to budge.

On March 24 1993, with commissioning now starting on site, TML wrote to Eurotunnel to say it was no longer willing to continue the negotiations. There would be no peace deal and no commissioning programme. TML said it did not know now when the tunnel would open and that it was now willing to refer the whole matter to arbitration, and would have the contract dispute panel set prices for the fixed equipment.

One TML source said: "We know this will take ages, but we think it will give us a better deal than if we accept what Eurotunnel is now offering."

TML's move was a brave one. It was clear that soon the contractor would have to start paying out its own money to fund the project, a prospect that scared some contractors more than others. None of the British five was faring well in a tough home market. The French contractors too were struggling although a more vigorous public expenditure programme gave them more of a commercial cushion than their British counterparts.

TML began detailed discussions at senior level to decide how to reduce site costs and keep spending down while it waited to fight its claim through the courts. And then, just five days after TML sent its letter cancelling further negotiations, another extraordinary row flared between the two sides. The International Chamber of Commerce gave its ruling over the principles of the dispute panel's powers to fix the costs of parts of the work and TML's right to pursue a global claim. Eurotunnel immediately issued a statement describing the ruling as "a very satisfactory outcome. A tribunal of the ICC has found in favour of Eurotunnel on all the substantive issues it was asked to consider."

Eurotunnel said that the tribunal ruled that:

1. TML cannot pursue a global claim for additional payments for the

(Top) The Shakespeare Cliff construction site, where the UK tunnelling machines began their journey, grew as the project developed. Spoil from the tunnel was used to turn a tiny strip of land into a spacious storage facility for TML.
(Middle) Construction workers used special trains to fit pipework in the tunnel.
(Bottom) The frontier between Britain and France is officially marked in November 1992.

The crossover chambers will allow trains to switch tunnels for safety and maintenance purposes.

Eurotunnel staff practise loading a shuttle train.

(Above) Le Shuttle is decorated in the corporate colours of green and blue.
(Below) A shuttle train being tested at the French terminal in September 1993.

fixed equipment work because its request to alter the payment basis from lump sum, as defined in the contract, to a cost plus fee was rejected. This meant, claimed Eurotunnel, that TML now had to submit all claims for compensation of the fixed equipment works on an individual basis, backed up with full documentation to prove the cause, the responsibility and its cost. In short, TML had to do just what it feared most.

2. The panel, which was in the process of fixing prices for the fixed equipment work, had to take into account the first part of the ruling in fixing these prices, and give both parties time to reformulate their positions on the fixed equipment to reflect those principles.

3. TML was not entitled to present any claim with respect to existing or future disputes for issues already settled by the 1989 Eurotunnel and TML Joint Accord. The Joint Accord settled principles for establishing the costs of tunnelling which could still emerge in theory in the future, and largely ignored the fixed equipment element of the project.

Eurotunnel was thrilled with the ruling and said that it "threw out the notion that TML's costs should decide its entitlements."

When it released its interpretation of the ruling to the public, its shares rose while those of the contractors fell. But a furious TML hurriedly issued a press release of its own – a strange event in itself, but by now the gagging clause in the contract was now largely ignored by the contractor. In its statement TML strongly disagreed with Eurotunnel's interpretation of the ICC ruling: "The news release issued by Eurotunnel today substantially misrepresents the findings of the arbitral tribunal," it said. "The tribunal of the International Chamber of Commerce has not found in favour of Eurotunnel on all the substantive issues, as alleged by ET. The tribunal did not accept Eurotunnel's demand that TML's fixed equipment claim should be dismissed. In particular it is misleading to state that TML must now submit its claims to Eurotunnel as set out in Eurotunnel's statement. On the contrary the decision of the arbitrators positively encourages the continued pursuit of TML's claims before the panel of experts."

The statement went on to say that the tribunal endorsed TML's right to pursue its claims and compensation for Eurotunnel's failings, and that the global and other effects of these "failings" will have to be taken into account in assessing the money TML should get.

The TML statement embarrassed Eurotunnel, which was concerned that shareholders would believe that they were deliberately being misled. Morton immediately moved to conduct a series of briefings with journalists to counter TML's interpretation. In an interview with *Construction News* he said that Eurotunnel was not asking for TML's claim to be dismissed, and had never said that that was the case. And, he said, although it was the case that TML could indeed go to the panel to settle claims, the ruling said the claims must be submitted to Eurotunnel first. Morton said: "We can't see anything at all in the tribunal's ruling which is against us.

TML must now present an itemised account of how they arrive at each claim – they said they did not have the documentation but now they have to show it to us."

Morton, now riled by TML's accusation that Eurotunnel was misleading the public, also used the interview to let TML know where the client stood if there was any deliberate slowdown on site. He said: "If TML fails then they're off. They have not failed yet, but some of the comments have suggested they may slow or stop work. Work is largely complete on the project and there is only the commissioning to do before opening. Anyone can do the commissioning."

Importantly, TML remained confident that it could still use the panel to fix its costs and submit a global claim. One TML source said: "Submitting claims one by one will take ages. Eurotunnel needs a deal soon if they want the tunnel finished quickly."

Eurotunnel was equally confident that a global claim would not be accepted by the arbitrators. The client's only fear now was that the work would be slowed so much that the banks would reopen the debate about funding. Morton was trying to stop any slowdown by issuing an "I'll sack TML" threat, but already rumours were emerging from the site that an element of non-cooperation had begun in the commissioning phase. TML was now not handing over paperwork concerning the multitude of system tests required on the scheme, preventing Eurotunnel from satisfying its own commissioning requirements.

Eurotunnel also revealed that without a deal, TML would start paying penalties in the form of liquidated damages from August 15 1993 which – the client said – would cost £240 million in the first year. On that basis, said Morton, TML's time blackmail did not work any more.

The battle to get public sympathy that was being waged by the two sides revealed itself again when Taylor Woodrow announced its annual results. The contractor's chairman Colin Parsons accused Eurotunnel of "intransigence". In an outspoken moment he said: "There is a very difficult owner [of the tunnel] who frankly doesn't have the money to pay for what he is committed to, and you have 10 contractors who are desperate to get paid. He asked us to build the contract. We have done it. Trains will be running through and people will be waving. All we want to do is get paid."

Eurotunnel and TML were now so far apart that even if a deal was possible there was no mechanism to get the two sides to be polite to each other for long enough to get signatures down on paper. It did not seem possible that things could get any worse. They did.

The specification changes ordered by Eurotunnel, and which gave rise to the whole fixed equipment row in the first place, also had a huge effect on the shuttle manufacturers who also had thought they were supplying "formica for the kitchen not marble". The shuttles had been delayed, with design changes being made even while the wagons were half finished.

Chief among these changes was the increase in the width of the fire doors of the passenger wagons ordered by the Channel Tunnel Safety Authority. These shuttles were being built in Belgium by ESC Wagons, part of Canadian firm Bombardier, which had a contract with TML – not Eurotunnel. The changes meant manufacturing costs had soared, as had the delay in supplying the 254 railway cars.

Just as TML knew Eurotunnel needed speed on site, so Bombardier realised that the best way of ensuring payment of the extra costs was to slow down production and delivery of the shuttles. Bombardier halted work at its factory in March 1993 and announced that the earliest any wagons would be delivered would be January 1994 – despite the fact that some were already packed ready for shipping.

Without the wagons, commissioning would be further held up and the passenger service was in danger of being forced to open in late 1994 – a date totally unacceptable to Eurotunnel and its bankers.

It was a new kind of problem for Eurotunnel. Although the dispute directly affected its ability to open the tunnel, because it had no contract with Bombardier it had limited scope for doing anything about the problem except put pressure on TML.

Although TML did not need the additional headache of a dispute with Bombardier, it was not about to strive to ease Eurotunnel's concerns.

Bombardier was complaining about design changes made by TML, but these were in fact design changes ordered by Eurotunnel. Indeed, they were some of the very same variations that had sparked the whole sorry fixed equipment dispute.

TML was able to tell Eurotunnel that it was increasingly unlikely that the tunnel would open for many months because of the design changes now directly affecting major subcontractors.

Jean-Claude Jammes – chairman of French member contractor SAE – warned that the dispute on site and the rolling stock problems would mean the project opening "sometime in 1994". But Eurotunnel dismissed his remarks as contractual posturing and repeated the view that the tunnel could still open in December 1993. This was now so unlikely that Eurotunnel's credibility was being undermined every time the date was repeated.

Eurotunnel's annual report in 1993, while taking a bullish view of the project as a whole, also betrayed some of the worries surrounding the project. In a letter to shareholders the joint chairmen Bénard and Morton said: "Our annual report this year makes something very clear. This magnificent project ought to open within a year and will represent a triumph of the determination, engineering skill and cooperation of many parties, including our bankers."

Stirring stuff, but the "ought to open" line revealed the fact that once again there was no opening date to offer shareholders. In fact, the report revealed that TML had submitted £912 million of claims, 97 per cent of which were for the fixed equipment. This meant TML was asking for around £1.5 billion total for this part of the work (£885 million extra on top of the original price of £620 million) of which, Eurotunnel believed, the contractor could only justify around £900 million.

Although TML had also submitted a 12-month extension of time claim, Morton said it made little difference to Eurotunnel if TML delayed a few more months. "TML is due to start paying penalties soon. The revenue we will lose by any extra delay is not as big as the money we are saving from not paying TML over the odds and the damages we will claim from them for being late. As to opening, June 1994 I think is highly unlikely and December 1993 is pretty unlikely. February is in the most likely zone, but it is undoubtedly not exactly right. TML is simply using the blackmail of time by delaying things. They have got to get their act together and decide who runs TML. We have a situation where the weapon of delay has become rather blunt in the hands of the party wielding it."

Sir Alastair Morton was as direct as usual. He was keen to get the message across that TML could not scare Eurotunnel with time threats. His justification for this, that the revenue lost would be lower than the penalties gained, was correct up to a point. But other factors were being conveniently ignored. TML was, for instance, capable of pushing opening into late 1994, so Eurotunnel would miss another chunk of summer revenues. And the banks were unlikely to be as patient as Eurotunnel.

The banks still watched progress carefully and demanded assessments from Eurotunnel all the time on how much the scheme would cost, what funds were still required and when they were going to start seeing some of their loans being repaid.

That spring's annual report showed that the total cost incurred by Eurotunnel up to a February 1994 opening date was projected to be around £8.4 billion at 1985 prices, up £300 million in the past six months. But Eurotunnel no longer believed cost to completion was the prime source of uncertainty on the project, despite the TML row. Instead, loss of revenue due to opening delays and the level of future interest payments were becoming the two key elements in calculating Eurotunnel's true financial position.

This figure was estimated by calculating the point after tunnel opening when Eurotunnel would start to break even – when revenues matched outgoings. After that date money would be coming in to start paying off the massive debt. So the funds required up to break-even point – the £8.4 billion to completion plus that required to keep Eurotunnel going until revenues built up – was the maximum funding requirement and was the key figure now to Eurotunnel and the banks. And this figure was dependent on revenues and interest rates.

Eurotunnel had already revealed that the delays had reduced estimated

1994 revenues from the £504 million assessed in 1992 to £260 million now. Very little of this would come in the early part of 1994, so Eurotunnel could withstand any delays by TML that affected the early part of the year. So Morton was able to use this to taunt TML that any delays would hurt the contractor in penalty payments more than it would hurt Eurotunnel in lost revenues. But if TML delayed into the second half of 1994, the effect would be more serious, a point noted in the report but not dwelt upon by Morton in his presentation.

As for interest rates, Eurotunnel would have borrowed £7.276 billion by the second or third quarter of 1994. Only £1.476 billion of this was allowed by the banks to be drawn down at fixed interest rates. The other £5.8 billion was therefore subject to the vagaries of interest rate fluctuations. Eurotunnel was keen to get these loans onto fixed interest. Every 1 per cent rise in interest rates would cost it £58 million a year, adding £350 million to its maximum debt in five years. Not surprisingly, Eurotunnel was talking to the banks about putting this money onto fixed rates as a matter of urgency.

Allowing for all the uncertainties, Morton said that Eurotunnel estimated that the break-even point would be sometime in 1997 and the maximum funding required would be £9.75 billion. Eurotunnel therefore estimated it had to raise another £750 to £850 million to see it through to break-even, and this would be done by a mixture of equity and debt. A rights issue was expected in 1994 or 1995, but basically this would be delayed as long as possible to allow revenues to build to boost confidence.

The long-term outlook, said Morton, was still very positive. "The tunnel's going to be immensely profitable for the next half century, but it will have an interesting five-year opening period to 1998," he said.

With Eurotunnel still needing extra funds to keep going, and with an almost mutinous contractor on site this was something of an understatement.

Eurotunnel desperately needed the tunnel open in the spring of 1994. Then it would see the revenues come in, could put the troubles with TML out of its mind, get the banks off its back and keep the Queen happy. She, after all, had now agreed to formally open the Channel Tunnel with French President François Mitterand on May 6 1994. Eurotunnel could do without the embarrassment of having two heads of state open the tunnel with no train to run through it.

21 Hail to the chief

"TML wanted a tough guy and a tough guy was what they got."

A project source

There are very few people in the world who have run multi-billion pound construction contracts and even fewer who have run huge tunnelling jobs. But American contractor Jack Lemley is one. And now he can add The Channel Tunnel to his curriculum vitae. Sir Alastair Morton was the man in the media spotlight and was the man who kept the banks and the shareholders reasonably happy. But although Lemley would never say so himself (he is quick to give credit to the TML team as a whole), it is the TML chief executive who can take much of the credit for construction being completed.

Yes, the tunnel was late and over budget and there were tragedies. But one of the greatest civil engineering projects of the century was handed over to the client on December 10 1993.

Among the five British and five French contractors that make up Channel Tunnel contractor Transmanche Link, hardly a harsh word has been said about Jack Lemley. He is hugely respected. As one project insider observed: "He even won the respect of the French contractors. And that is a rare achievement." But for those outside TML, dealing with Lemley could be a difficult experience. "TML wanted a tough guy and a tough guy was what they got," said one source close to the project. "They needed a battering ram not a feather duster. I have a lot of time for him but there were times when he was possibly too tough and not as diplomatic as he might have been."

Although he seemed laid back, he had a short fuse if things went wrong. "He was mercurial, certainly. But his was a difficult task. To appease both Sir Alastair Morton and the 10 TML contractors would have required the patience of a saint."

Jack Lemley returned to the United States just before Christmas 1993 for, as he put it, "a little rest". After four and a half years at the helm of TML, even his most fervent enemies would have admitted that he deserved it.

Lemley, one of a number leading American contractors who occupied senior positions on the project, arrived with the reputation of being a somewhat taciturn but strong-willed individual, and sources say he clashed heatedly and frequently with Eurotunnel bosses Morton and John Neerhout. This may explain why Eurotunnel declined to comment on Lemley's departure or his achievements as TML chief executive. Asked at the time by *Construction News* for a reaction from Sir Alastair, a Eurotunnel spokeswoman would only say: "We wish Mr Bedelian (his successor) all the best."

Yet despite the huge stresses and strains, Lemley insists he has enjoyed the job. And he is leaving the Channel Tunnel with a huge respect for the quality of the two countries' construction industries. "It was a fabulous job, there is no doubt about it. We had so much animosity and so many differences over the contract that a lot of the fun did go out of the job. But from an engineering construction point of view, it was a fabulous project. There was a sense of history felt by even the most sceptical of the mining crews." However, he stressed: "It was unusually hard work. It was such an enormous job physically, it took a tremendous amount of effort to manage the project, to ensure we had the right resources at the right time. We have had plenty of bruises and problems but we kept the main mission in sight."

His praise of the British and French workforces will startle those more used to the bad press the industry receives from commentators this side of the Atlantic. "It is hard to imagine having a better team to do the job with. We probably got the best performance out of each individual than I could have hoped for."

At its peak, the project employed 15,000 people, yet there was not a single major industrial dispute. "I am very proud of the teams on both coasts. I have never seen labour workforce work harder. I have an enormous respect for what they have done. They are well-educated, respond to training and produced outstanding results."

He added: "I dealt regularly with the trade union leaders in Britain, and to a person they were fully behind the job – they bent over backwards to get the job done. I think they recognised that the project was important as a testament to what the construction industry could achieve."

The high point was undoubtedly the first breakthrough under the Channel on December 1 1990. "I have an immense pride in what our team had accomplished. We had physically joined the two countries."

A single low point is more difficult to define. "I never thought the project would not finish," Lemley said, "and there was not one single low point, but several – the fatalities. They were terrible tragedies. If there was one thing I could change now it would be to put more effort into the safety aspects. We had a good safety programme, but construction as an industry has a long way to go both in Europe and the USA."

Lemley, who was 58 when he stepped down from TML, is a former president and chief executive of the Blount Construction Group, spent 10 years working for Morrison-Knudsen (where he was responsible for all the firm's heavy civil engineering projects worldwide), and before joining TML he was running his own consultancy business.

Projects he worked on included the King Khaled military city in Saudi Arabia which in the early 1980s employed over 12,000 construction personnel; a vast US$2.25 billion (£1.5 billion) opencast mining scheme in Columbia (this included a railway, two airfields and a port, and came in six months early and 10 per cent under budget); and in the mid 1970s, a US$600 million (£400 million), 13.5 mile water main supply tunnel 800 feet below New York City.

On this last job, he was project manager for contractor Guy Atkinson, a firm involved in the Channel Tunnel scheme abandoned by the Labour Government in 1975. "I remember sending several of my people to the Channel Tunnel then," Lemley recalls.

One of those instrumental in appointing him to lead construction of the Channel Tunnel was Sir Frank Gibb, then chairman and chief executive of Taylor Woodrow, one of the 10 British and French firms making up TML. "We ended up interviewing three candidates and we all agreed that Jack Lemley was the best man for the job and there is no doubt that he was," Sir Frank said. "He is quiet and thoughtful, he does not jump to conclusions and he does not get involved in enormous rows. But he does take a firm line, he is a good leader and he is very experienced in handling contracts."

Lemley himself added: "I think I brought to the job a certain management style, a quiet decision making process that allowed the project to move forward. We tried to empower the people on the project to make decisions and to make them accountable, but didn't crucify them if something went wrong."

Lemley stresses that the huge scale of the project must always be kept in mind. This was no run-of-the-mill civil engineering contract. "Some of our managers had more spending power than some secretaries of state." But having reached a decision, he does stand his ground, as Eurotunnel discovered, and Lemley conceded: "I do have a short fuse at times, a very short fuse, and there were times when there were blistering rows. But I do not accept all the blame for that. I was dealing with some very difficult personalities on the other side." Of his relations with Sir Alastair Morton and Eurotunnel, Lemley is remarkably sanguine, considering the number of rows the two men are rumoured to have had.

A great percentage of the bad feeling between the two sides simply revolved around the price of the work. TML believed Eurotunnel was demanding improvements not covered by the contract, especially under the lump sum section of the contract, but was refusing to pay for them.

Eurotunnel believed it was holding TML to a contract that the original contracting members of the Channel Tunnel Group and France-Manche, the winners of the mandate back in 1986, had been instrumental in drafting. "We both had a job to do," Lemley simply said with a verbal shrug, And he conceded that for all their differences, Morton is clearly a "brilliant man...with an enormous capacity for detail and for handling difficult situations. I can't fault the energy and effort he put into the project."

A more technical area of concern was the difference in design standards between Britain and France which, Lemley said, meant that structures such as bridges cost more on the British side of the Channel. He declined to say that such structures are over-designed in the UK. But he added: "The French are less robust but more realistic from a cost standpoint. For example, look at the standards of over-bridges in the two terminals where one used French Government standards and the other UK Department of Transport standards. There are significant elements in the UK design work covering risks that the French do not include."

But it is the antagonism between client and contractor that clearly causes Lemley most regret and which dominates so much of the debate about the project. Many of those involved feel the disputes themselves caused delays as both sides focused on how to defeat the other rather than on completing the project.

"I did not enjoy the differences with the client. I felt the job could have been done more efficiently and smoothly if we created a less adversarial relationship," Lemley says. "Make no mistake, construction went very well. But with better client relationships, it could have been improved." The rows, he reckoned, cost as much as £1 billion and caused the job to be completed between nine months and a year late.

Yet despite the problems and Eurotunnel's complaints about the contractors, TML succeeded in setting standards "that have never been met before. What Eurotunnel has there is first class in every respect. There has been no scraping on quality".

But if Lemley is critical of Eurotunnel and how it handled relations with TML, he is, if anything, even more fiercely critical of the British and French governments and their role in causing delays and problems on the project.

The problem revolved around the Intergovernmental Commission which had to approve the project as it was designed and built – but it was taking too long to reach its conclusions. The result was that various parts of the work had to be reworked after completion – for example, widening some of the doors on the shuttle trains – at great cost.

"The governments have an enormous responsibility here. They created an environment that was not easy to work in and I think they let the concessionaires down badly."

For all the disputes and concerns, Jack Lemley is proud of what he achieved at TML. But he certainly sounds more relaxed now that he is back on the other side of the pond at his home in Boise, Idaho.

22 Site warfare breaks out

"It is in the lap of the Gods when the tunnel will open."
A TML contractor

Eurotunnel's letter to its shareholders in April 1993, written jointly by chairmen Sir Alastair Morton and André Bénard, included a small paragraph on the relationship between the staff of Eurotunnel and TML. It was a fairly unremarkable statement, following as it did a now conventional moan at TML for using the opening date as a contractual weapon and a demand that TML cooperate with its client and design, supply, construct, install, commission and guarantee the performance of the scheme as contractually bound. The letter then said: "Fortunately, at technical level there is good cooperation between Eurotunnel staff and TML staff and suppliers, when not fettered by contractual manoeuvres." This point was underlined by Morton in his London press conference. He went out of his way to stress that at site level the two teams were getting on just great. The problem, apparently, was the levels of management above them.

Morton used the word "blackmail" to describe TML's actions regarding site progress – an accusation that angered TML. But what Morton did not say at that stage was just how bad the site problems had become and the extent of disruption that had developed in Kent.

With all good will removed on both sides, petty bickering had quickly turned into petty acts of disruption which had reduced site operations to farce. If it was true that the site teams just wanted to get on with the job and were getting on well together, as described in the chairmen's letter, then they must have been horrified at the actions of their management that were dragging the greatest civil engineering project in Europe through the mire.

As it had become clearer that TML and Eurotunnel would not do a deal on the fixed equipment costs, then TML had begun to disrupt site activities. In a campaign that could only be described as site warfare, TML used every weapon at its disposal to make life difficult for Eurotunnel. When the client wanted to carry out VIP visits to the site, it had to get permission to do so from the contractor. This is normal, as the contractor runs the site. But TML had taken to cancelling these visits at the last minute, embarrassing its client in front of bankers, engineers and politicians. Eurotunnel would bring people to visit the work, only to have TML say at the last minute that there was no access to the site at present and waste everyone's time and effort.

At the terminal some buildings had already been handed over to the client for their commissioning and testing work. But when Eurotunnel sent their staff in they would find that the locks had been changed and they were unable to get into the buildings.

Alternatively, TML would cut the power supply to the buildings that had been handed over, making Eurotunnel's work impossible. No-one had seen a dispute escalate to this level before. The construction industry has something of an unfair reputation for building late and over budget. Despite great improvements in the industry's performance, most of the people connected with the project had worked before on schemes where there had been local difficulties. But no-one had seen anything like this. And, worst of all, it was happening on the industry's showpiece prestige scheme.

Eurotunnel described the disruption as "TML playing silly buggers", but the client did not emerge from the fiasco with much credit either. It was also being awkward, and in doing so was choking TML's cash line. TML was being paid for each test it carried out on the project to the satisfaction of the client. Under the letter of the contract it was not paid for the test until all the paperwork related to that part of the work was handed over to the client.

This involved getting together a mass of documentation for each test area, to prove the quality of the materials and workmanship in that phase of the work.

On a normal contract, with a bit of give and take on both sides, the contractor would expect to be paid once the test was done and checked by the client, and the paperwork would follow later. By now this was no normal contract. Eurotunnel was demanding every last bit of paperwork before it released the money to TML. Eurotunnel knew that TML just did not have this documentation, for two reasons. Firstly, the site teams had been concentrating on getting the work done as quickly as possible to catch up time on the project, especially when it seemed a deal would be struck and work would have to be completed in December that year. The work was being done at the expense of the administration. But the second reason for there being no back-up documents was that the TML administration staff had all been ordered to work on the details of the fixed equipment claim, leaving too few people to compile the day to day quality management information for forwarding to the client.

Eurotunnel knew that unless TML directed resources away from the claim and back onto the job in hand, it could squeeze the contractor financially.

On a project which people were proud to work on, this was not a time either side could look back on with any satisfaction. As Morton said in April 1993, the actual site workers were getting on fine. They looked on in horror while their bosses apparently did everything possible to scupper their work.

TML was still maintaining at the time that it was not party to any "blackmail" and that it was working to complete the project as quickly as possible. The site disruption it put down to genuine requirements to switch off electricity or keep visitors away at difficult times, as would be expected on a project of such magnitude and against such a tight schedule.

TML joint chairman Joe Dwyer issued a statement condemning Morton's blackmail charge as "astounding nonsense", and claiming that work was proceeding quickly and effectively. "This is just Alastair being Alastair," he said. The Dwyer statement read: "In the absence of a commercial agreement (of the fixed equipment dispute)...TML will nevertheless proceed, within the terms of the contract, to achieve completion of the project at the earliest possible time."

Dwyer said that the delays were well documented and were down to changes in the performance and safety specifications and that TML would wait for the panel to rule on the monies that were due to them. And when it did, this would be paid immediately and in cash – not shares or securities.

Despite its brave face, TML was starting to suffer financially on site. The 10 contractors met and agreed each to start paying out around £2 million of their own money each month to keep the project going. But this figure was due to rise quickly to nearer £10 million a month, money the contractors could little afford.

The pressure once again began to show in public bickering. One TML executive, angered by Morton's latest tirade against the contractor, was asked when he thought the tunnel would open. "The only thing that will be open for certain is Morton's mouth," he spat in reply.

The anti-Morton lobby again rallied and called for the man's head on a plate, following a string of letters written to the contractor slamming its performance and demanding better progress.

TML chiefs told the *Sunday Times*: "Only with Morton out of the way and a more reasonable man in his place can this project be finished. His behaviour is disgusting. If you want to know what the main obstacle to this project is, it is him."

Morton, for his part, laughed at suggestions that he was being obstructive and offered to swear on "a stack of Bibles" that it was the other way round. "We were the ones who initiated the peace talks and came within a whisker of achieving them. But they walked out. It is us that went back repeatedly with the new ideas, not them," he said.

But with the City now openly discussing who would take over the project once Eurotunnel was forced under by the dispute, there was little doubt that the pressure was growing even more.

The banks were still backing Morton and his team, but were once again dismayed by the level of disruption on site and the potential consequences. They were surprised by the extent both sides were willing to go to in order to disrupt the other and were becoming increasingly of the opinion that the idea of leaving the two sides to fight the claims through the courts while the tunnel opened for business was becoming pie in the sky. TML was clearly not going to acquiesce that easily. While the tunnel was still shut, it still had power.

As if to underline this case, TML played its next card. With money getting increasingly tight for the contractor, it finally carried out the threat that it had been hanging over Eurotunnel since the beginning of 1991. In a move that sent shivers of fear through Eurotunnel and its banks, TML slashed its work rate on site by a massive 75 per cent.

Money was becoming extremely tight for the contractors. Eurotunnel's squeeze on TML was totally effective and the 10 member companies were starting to pour money into the project to keep work going. Despite having threatened to slow down work or even stop altogether several times, this was never seen as a totally serious option and certainly a weapon of last resort. But all of a sudden the contractors found themselves funding work on site to build a scheme for a client that, as far as they were concerned, would not pay them for their work. Worse still, they were operating an accelerated programme which was costing a fortune. This time the contractors decided to bite the bullet: subcontractors on site in Folkestone were told to cut their work rate by up to three-quarters in a bid to save TML money. From working three eight-hour shifts a day, seven days a week in the spring to hit tight deadlines, the UK commissioning and testing subcontractors were told to work a single shift, five days a week. Instead of 168 hours of work a week, the hours were cut to just 40. And in France, where two ten-hour shifts had been the norm, the pace was halved.

A senior Eurotunnel executive called the move "tactical" but agreed it was having a marked impact on the site. He told *Construction News*: "TML is now funding the work itself, which must be a strain. This move to single shift is tactical – it means they can work as inexpensively as possible but of course work will take longer."

Eurotunnel was working out how it could get round the problem, he said. "There has been an impact on the project from TML's move. We are trying to work round it and we will not change our opening date until the time lost is totally irrecoverable. We will keep the same staged opening dates but maybe with a more reduced service than we hoped." The move was TML's biggest gamble to date. With the deadline for liquidated damages due in August 1993, the contractor was playing its last card and challenging Morton's statement that the time weapon was now blunt because of the profile of the revenue build-up.

Even with the accelerated programme work was not going to finish on site for at least six more months, so by cutting the pace so dramatically TML was throwing up the possibility of work dragging on for more than another year, putting the summer 1994 revenues in serious doubt.

TML was also still digging in its heels about a phased opening, which it had never agreed to work towards, further threatening Eurotunnel's profit stream. And with shuttle delays still holding up the commissioning and Bombardier refusing to deliver wagons from Belgium, TML was preparing to argue that it could not be charged liquidated damages in August because the delay was not the fault of the construction team, but was down to the rolling stock delays.

TML backed up its moves on site by maintaining the line, at least in public, that the tunnel could not open for a year. TML vice-president Philippe Montagner told shareholders in May 1993 that although building work on the tunnel was virtually complete, it would take more than a year to test the equipment and systems and the tunnel might not open until the end of 1994. He said: "The installations as such are on the verge of being finished but that is not the case for software or tests, and in the best of cases another year of work is still needed."

This line was backed up by every TML official who was asked about opening. The typical response was: "It is in the lap of the gods when the tunnel will open. Its opening depends on cooperation between TML and Eurotunnel and clearly we have not got that at the moment." All Eurotunnel had to do, said TML, if it wanted an opening date, was to come to the table with a new offer.

TML did not publicise the site disruption, preferring to keep that under its hat to maintain as much public sympathy as possible.

While TML played its time card on site, plunging the project into dangerous disarray, a change was being made to its management structure which would soon be hailed as a key turning point in the project. Neville Simms, chief executive of Tarmac Construction, replaced Joe Dwyer of Wimpey as head of the UK arm of TML. Dwyer was hugely respected in the construction industry but his relations with Morton had soured. Simms was rated as a deal maker and a man who had time and respect for Morton – which was more than many of his colleagues could claim. One insider said: "If anyone can do a deal with Eurotunnel, Neville Simms can." Simms himself said at his appointment: "It is important that the developer and the builder walk hand in hand for the last mile." Simms immediately started actively seeking meetings with Morton to bring the project back on line.

Why exactly Simms succeeded where everyone else had failed is not clear. As with all such situations, it is often a combination of factors which conspire to throw light on a subject where there had been none before.

Without doubt, TML's site tactics had been effective and were applying pressure to Eurotunnel, so much so that one senior site source said it was the main reason why Eurotunnel signed a deal.

But added to that was the fresh face factor, and the fact that that fresh face was Simms – a man used to doing deals and a man good with dealing with Morton.

A quiet, laid-back man and only 14 months into his position as chief executive at Tarmac, Simms had rarely had personal run-ins with anybody at Eurotunnel. Even when dismayed with progress or breakdowns in negotiations, Simms was not a man to queer his pitch by shouting and ranting or complaining to the press. Most important, Morton seemed to feel that Simms was a man he could deal with. But added to that was a third factor – the intervention of the Bank of England, which now stepped in to broker a deal. It was the second time the Old Lady of Threadneedle Street had helped in this way. This time it seemed it would be brokering the deal which would result in the completion of the tunnel.

In just 10 meetings, five refereed by the Bank of England, a solution to the crisis was found. Out of the blue Eurotunnel and TML called a press conference together on Tuesday July 27 to tell a rather surprised world that a deal had been struck that would see the tunnel open in March 1994. TML had agreed to hit a series of deadlines over the next four and a half months to hand over the project to Eurotunnel on December 10. In return, Eurotunnel would pay TML an advance of £235 million, so the contractor would no longer be out of pocket.

From December to March Eurotunnel would be in charge of having the final testing programme completed. TML would be involved in the work but would be paid separately. The phased opening would start in March 1994 and build up to peak in 1995.

The £235 million, plus £200 million still held by TML on Eurotunnel's account from the time of the panel's extra funding award in 1991, would subsequently be returned to Eurotunnel unless TML could justify claims to cover the money. It was, not surprisingly, a simple deal. And although it was by no means comprehensive – the fixed equipment claim was deferred rather than settled – in one go it solved the contractor's funding crisis and pretty much guaranteed revenues starting for Eurotunnel in March.

It was what Morton called a "win-win" agreement. "It is not meant to be comprehensive," he said. "But it takes us from here to opening and sorts out any time claims. Other financial matters will be sorted out under the contract next year."

Simms added for TML: "This agreement has taken the smoke out of our eyes and allows us to get on with it. But it does not take away our right to claim for extra payment."

The Channel Tunnel had seen some of the most bitter rows and disruptive actions ever to hit a construction scheme. It had been taken to the verge of collapse many times. There had been accusations of misrepresentations, blackmail and intransigence. No-one had seen anything quite like it before. But after all the rows, the remarkable survivor Sir Alastair Morton and the unflappable peacemaker Neville Simms cracked open a bottle of champagne, touched glasses in a clink of mutual respect and stood shoulder to shoulder smiling for the cameras. The most amazing civil engineering project the world had seen was saved. The Channel Tunnel, after a wait of some 200 years, was actually going to open.

23 The slow train to France

"There is a real risk that Eurotunnel could be faced with a tunnel that is completed but cannot be used ..."

A TML director

One of the most remarkable aspects of the July 1993 deal between TML and Eurotunnel is that with the stroke of two signatures the tunnel crisis was successfully defused. Having bickered, battled and berated each other for so long, it seemed impossible that the two sides could complete the work. But as Morton had said in April, relations were good at site level. Indeed of the thousands of people employed on the project, most had got along very well no matter who their employer was and had developed fruitful working relationships. It was just a few key individuals who had fallen out and created a major dispute for the world to watch disapprovingly. So once Simms and Morton had made the deal, it was not that difficult to establish cooperation on site and hit the milestones that would lead to the hand-over in December.

Only one obvious problem arose, when the *Mail on Sunday* reported that there had been an explosion and massive electrical fire in the tunnel during testing which had wiped out a 500 metre length of cable. Workers were quoted as saying that if people had been in the tunnel at the time of the incident there could have been a tragedy. But TML moved quickly to calm the situation, and said that while there had been a fire during tests it was not a major event. The contractor said the fire was caused by the loss of a ceramic insulator which allowed live wires to touch their fittings and cause a short circuit.

The explanation was later confirmed by the Channel Tunnel Safety Authority, which was represented on site at the time, and in March 1994 the Press Complaints Commission upheld a complaint against the *Mail on Sunday* over the story. Apart from this minor incident work progressed well on site. Tunnel watchers always expected another row to emerge, but it never really did. Disagreements at management level about how to hold the hand-over party were observed with interest. But for a time TML was happy, receiving its money from Eurotunnel – albeit only "on loan" – while Eurotunnel was busy changing into an operating company capable of running what would eventually become the busiest stretch of railway in the world.

Both sides still had their fair share of normal contractual problems to sort out. But these were much representative of the typical frustrations of a large civil engineering job rather than the sort of dispute that could bring the whole industry into disrepute.

TML still had Bombardier to deal with. The Canadian company had been cajoled into restarting the production and delivery of its 254 passenger wagons and this was seen as a vital move if the manufacturer was going to make enough wagons available, for commissioning and subsequently for the phased opening of the project.

The company restarted work after striking a deal with TML on the technicality of the contractor accepting that extra work carried out by the manufacturer had been authorised and should be paid for. As with the main TML and Eurotunnel deal, this was an enabling agreement. Although the money would still have to be agreed, the important thing was to get the wagons on site and get the project running.

At one stage, in the recent past, Eurotunnel was seriously considering bringing in alternative rolling stock to start the services. This was not now necessary, and by October 1993 as many as 43 wagons were on site, allowing the assembly of the vital test train and eventual start-up of the passenger shuttle service. Bombardier's original claim for around £230 million had risen sharply to £388 million, and this on a contract that was originally only priced at £360 million.

For Eurotunnel there remain plenty of worries, particularly about future revenue. Its October interim report showed that just 15 of the 38 locomotives needed for the full service were on site, but that all 228 of the freight carrier wagons had arrived. These were due to enter service in March 1994, two months before the royal opening and the start of the passenger shuttles. And when the tunnel was successfully handed over in December 1993 and the train fares announced in January 1994, there was every reason to believe that the March date would be achieved.

But once again the sheer complexity of the final testing foiled Eurotunnel, and in February 1994 the company sheepishly announced that the HGV shuttles would not start running in March and a full passenger shuttle service would not start in May as planned.

The company had underestimated the time it would take thoroughly to test the complex state-of-the-art locomotives and this, coupled with the delay in the arrival of the locos on site, meant the start of service had to be put back if safe operation was to be assured.

EVENING STANDARD, 13 JUNE, 1990

"Will the passengers from the French High Speed Train arriving from Paris transfer to the number 11 tram for Kings Cross!"

Instead, the company hoped to start running the HGV trains around May and the passenger trains as soon as possible after that date although the official opening by the Queen and President Mitterrand would go ahead as planned on May 6, 1994.

In 1991 a TML boss at the height of the dispute with Eurotunnel said: "There is a real risk that Eurotunnel could be faced with a tunnel that is completed but cannot be used because not enough trains are available for testing and commissioning the system. We are not the problem on this project – the rolling stock is."

It was an inspired observation.

When the trains do start to roll, then at first two freight shuttles per hour will run, building quickly to three. The passenger shuttles will also build up during the summer and autumn and at peak there will eventually be four per hour. The Eurostar through train service between London, Paris and Brussels run by British Rail and SNCF was expected to begin in June 1994, but the high speed rail link between London and the tunnel is unlikely to be completed until the year 2000.

Before this latest delay Eurotunnel needed to raise another £1 billion to cover its maximum funding requirements of around £10 billion. This has risen since April because of the £235 million advance agreed in the July deal. Break-even is expected in 1998. Whether this figure will change with each further month of lost revenue – each of which will cost the company around £50 million – remains to be seen. But a rights issue, due for the summer of 1994, is intended to raise up to £750 million, with another £500 million coming from the 220-strong banking syndicate.

Meanwhile Eurotunnel has pursued claims against the British and French governments and against British Rail and SNCF.

Eurotunnel was claiming up to £200 million from the two governments for their failure to provide the infrastructure promised under the concession agreement, and for additional costs incurred by the developer due to safety changes. This claim was dropped at the end of 1993, when Eurotunnel was given an extra ten years on its concession, boosting the time it operates the tunnel from 55 to 65 years.

The claims against the railways for delays to the through train service are continuing and will now be settled over years of wrangling after the tunnel is open. As will TML's row with Bombardier and, of course, the famous fixed equipment claim.

At the end of December – as required under the July protocol – TML submitted details of its claim. The total now stood at £1.95 billion and both sides seemed ready to fight the dispute out through the courts. Just like the good old days.

Further delays with the commissioning process meant that the date for the start of full freight and passenger services remained uncertain and the financial pressure was growing prior to the rights issue. Eurotunnel needed to settle their dispute to give the issue a boost and TML hardly wanted to wait five more years for its money. In April 1994 the skill of Simms and Morton as negotiators was displayed for a second time as finally the four year fixed equipment battle was resolved. TML would be paid £1.14 billion – and all in cash. Although well short of its claim it was considerably more than it was offered in 1992 when the cash element was less than £1 billion.

TML would also underwrite £75 million of the rights issue, throwing its weight behind the project's future.

Both Simms and Morton liked the deal and suddenly the rows were over, solved seemingly with ease without pound by pound haggling or more legal fees.

On site the commissioning problems still threatened problems to come – Eurotunnel stressed that the deal did not include rolling stock procurement – but the April deal finally brought to a close the construction of the Channel Tunnel.

Epilogue

Can a scheme that has been completed a year late and at more than double the original estimated cost be considered a success? In the case of the Channel Tunnel the answer must be "yes".

The construction industry is not held in high regard by the British public. But it was a construction industry initiative that caught the imagination of the public and politicians on both sides of the Channel and enabled the dream of a fixed link between Britain and France to become reality.

From the very start the two governments stressed that no money nor guarantees would be made available to whichever consortium won the mandate to design, build and run the scheme. Private sector finance for infrastructure projects is an idea still struggling to get off the ground in the UK. Even the public/private initiatives now encouraged by the British Conservative party are only progressing at a snail's pace.

London's CrossRail project, the London Underground Jubilee Line Extension to Docklands and the high-speed rail link between London and the Channel Tunnel are three excellent examples of privately financed or public/private schemes which have been stuck on the starting blocks for years. And that is despite the fact that they are receiving at least some government money.

What Eurotunnel and Transmanche Link did – without help from the state – was build a scheme which will benefit Europe as a whole. And for that the contractors, Eurotunnel and the long suffering banks should all be praised.

But even though construction has been completed, Eurotunnel is still at the time of writing struggling with the commissioning process – making sure all the systems work as they should. This means it is still under great financial pressure. Every extra day that goes by without trains and passengers paying money to run through the tunnels is costing the developer dear. Another rights issue, due in the summer of 1994, will be a key test for the company and of public and City confidence.

Eurotunnel, though, has faced grave problems before. Above all else, it is a great survivor. Two or three times during construction it faced financial crises and overcame them. If the company had succumbed and the project had been thrown into disarray, then privately financed infrastructure projects would have been set back 10 years.

But the banks have kept faith. The public remained excited by the project and the drama surrounding it. From a construction viewpoint, the Channel Tunnel has taught everyone involved in the industry a great deal – about safety, organisation, and quality, for example – which will influence how work is carried out for decades to come.

The tunnelling, albeit through good ground, reached speeds not before thought possible by full-faced tunnel boring machines. A level of speed and efficiency was attained by the site trains charged with the disposal and delivery of muck and materials that will be an inspiration to other schemes – and the underground crossovers were construction achievements that will be remembered with huge pride by all who worked in them.

And nobody will in future underestimate the problems of fixing the mechanical and electrical systems into a tunnel and getting them all working satisfactorily. TML achieved this in a tunnel connecting two countries with two site teams who had different ideas of how to build, how to install and how things should work.

This was without doubt a huge achievement.

The project did however kill 10 men. But those 10 men worked hard to fulfil the dream of building a tunnel between England and France, and they would be rightly proud that despite all the wranglings and rows the tunnel was built.

The Channel Tunnel produced some of the most remarkable rows ever seen on a construction project. But its successful completion is a tribute to Eurotunnel's great financial skills, especially those of Sir Alastair Morton, and to the pioneering, persistence and perseverance of the client, the contractors and their workforces against all the odds.

At the time of writing, all that now remains is for Eurotunnel to overcome the final snags and to persuade the British and French public to use the tunnel. Despite the rows, everyone involved on the project is desperate for it to be an outstanding success. They believe it will be, and that the Channel Tunnel will transform our relationship with our European partners. The only question is, how long will it take?

Index